The Rhythm of Life and Death

The Rhythm of Life and Death

Avijit Pathak

THE RHYTHM OF LIFE AND DEATH
Avijit Pathak

First Published 2011
Reprinted December 2011
Reprinted 2014
Reprinted 2017
Reprinted 2022

ISBN 978-93-5002-156-9

Published by
AAKAR BOOKS
28 E Pocket IV, Mayur Vihar Phase I, Delhi 110 091
Phones : 011 2279 5505, 2279 5641
aakarbooks@gmail.com; www.aakarbooks.com

Composed by
Limited Colors, Delhi 110 092

Printed at
Sapra Brothers, Noida.

To all those who made me feel the
rhythm of connectedness

Contents

Preface

I thought that I would write yet another academic text. But then, I ended up doing something else: creating this small book of prayer and contemplation—a book revealing the rhythm of life and death. It consists of a series of brief pieces written at intense moments of pain and suffering, and love and revelation. Needless to add, there is no academic method, no design. The entire exercise is spontaneous and meditative.

It is for the reader to find her/his own meaning in the book. However, at this juncture, I wish to state that there are three moments of revelation which, I guess, have been reflected throughout the book. First, I have begun to value the virtue of calmness: the ability to remain stable even when there are surprises, inexplicable tragedies and storms in life. This is possible only if we accept—and accept with grace—the totality of life: its peaks and valleys, births and deaths, and successes and failures. This acceptance is not passivity. Instead, with this acceptance begins a more meaningful engagement with life. We acquire the courage to accept that the phenomenal world is inherently impermanent; and no matter how much we try, the empire we build up today would fade away tomorrow. And hence the task is not to withdraw, but to act, and yet remain a composed spectator of this play. Second, this quest for a meaning makes us realise the limitations of our egotistic existence. Our egos become overwhelmingly powerful when we begin to think that our physical existence is our primary substance, our bodies are our defining features, and life is just a pursuit of what this embodied existence demands: wealth, prosperity, fame, power and sexuality. This ego is limited—confined to the boundaries we have erected around ourselves. Not surprisingly, it remains insecure; it leads to fear—fear of losing, fear of nothingness, fear of death. And with fear begins violence and aggression. There is no freedom from fear, insecurity and violence unless we realise that this body, far from being our real substance, is only a carrier of unbounded energy—the energy that transcends all limits, and is flowing through everything. Living life meaningfully is not to get obsessed with our limited/egotistic existence, but to realise this energy, and allow it to manifest in all that we do. Of course, our bodies will disintegrate; but the energy would remain, be liberated from the container, and spread out. Seeing beyond the finitude of the body/ego is to experience the rhythm of connectedness with everything—the tree, the mountain, the river, the distant star. Love is this connectedness. Love conquers fear. Third, we ordinary mortals too have our sacred moments of awakening: when with absolute enchantment we find our poetry in the phenomenology of everyday life. What is ordinary becomes extraordinary; the finite expresses itself as a manifestation of the infinite; and through pain and suffering we purify ourselves, and enter the domain of possibilities.

You and I need not be able to arrive at this peak of awakening. We fail; we fall down. Yet, there are moments when we experience the flash of truth. I value these moments. And I begin to pray: let these moments continue to inspire us.

The book emerges out of my intense urge to involve my readers in this celebration of prayer.

April 2, 2011 Avijit Pathak
New Delhi

1

High Culture, Radiant Moon and the Tale of a Taxi Driver in the Metropolis

Delhi is huge and gorgeous. As a metropolis it has its vibrancy and multiple colours. The very rationale of a big city is being experienced every moment. Its expanded roads and flyovers, the continual flow of ever increasing vehicles, the colourful banners demonstrating the visibility of the corporate world, brand names and their associated mythologies, the fortified apartments creating an exclusive world of their own, and the spatial/architectural design of the malls and multiplexes do indicate that the city is fast becoming global.But then, the city, despite the appearance of pleasure and excitement all around, has its own contradictions. It has its margins and innumerable narratives of humiliation, oppression and violence. There is poverty amidst consumption, violence in seduction, and there is perpetual restlessness emanating from desire, competition, fashion, deprivation and survival anxiety. Poetry dies; utopias disappear; and the mind gets deserted.

However, there is a call that I often hear: Don't give up, strive for beauty, for all that is simple, natural and authentic, free from the armour, complexity, packaging and violence of the metropolis. Is there something in this huge city that can assure me that life, despite all odds, has its softness and inspiring moments? Possibly, as I thought, it lies in the domain of art. Yes, the city has its own space for the creation, production and consumption of high art. Its art galleries, theatre halls, specific book shops and coffee homes do attract people.

So one day I decided to see a play—a play that derived its storyline from Henrik Ibsen. Ibsen, we all know, was radical; and the city, I felt, accommodates its radical intelligentsia. They assemble whenever there is a film/theatre festival; they come together if there is violation of human rights in Jammu and Kashmir or in the North East;

and they are seen at book release functions or seminars on socially relevant themes, be it postcolonialism or gender discourses or global warming. And hence in the theatre hall I found myself amidst them. But far from experiencing joy, beauty and simplicity, I had a sense of loss. Even radicalism, I felt, was filled with the elitist smell of cultural capital and intellectual snobbery.

Meanwhile, the play began. No, I did not like it. Yes, I love to read Ibsen. But the way he was used in the play was not aesthetically soothing. It lacked substance and depth. Forms and technologies became overwhelmingly powerful, and Ibsen was reduced into a spectacle—artificial and ornamental. The artificiality of the city was reinforced again. Needless to add, I was tired and exhausted. I failed to find my art and aesthetics—my moments of purity and calmness.

'Papa, look at the moon', my daughter awakened me. Indeed, as we came out of the drama of high culture, the radiant moon above the skyscrapers was waiting to welcome us. It was extraordinarily illuminating. The metropolis with all its wealth, technology, speed, mobility and

pretence might not have the time to notice it. However, like a piece of eternal truth it was there, and it would always be there in silence and with all its glory. The radiant moon made me feel once again—aesthetics is an experience of a deep communion with all that is simple, natural, rhythmic and beautiful. Indeed, as I looked at the moon in the downtown of the metropolis, I began to pray: 'Give me the strength to find you all the time. Protect my way of seeing and experiencing. I should not lose it even when the elitism of high culture, the burden of radicalism, and the armour of city life tend to disrupt the aesthetics of our inner world'.

The taxi was waiting for us. What else could we do? Here is a city that has systematically destroyed its public transportation system. Private cars with their ever changing models have begun to symbolise the aspirations of the new middle class. Seldom does the global city in the neo-liberal era bother about shared public facilities. The city excludes those who cannot cope with its logic of development. If you do not have a car, keep waiting for over- crowded buses—and that too with a terribly low frequency. I, however, hired a taxi. Mr. Pal Singh—a Sikh gentleman with a compassionate face—drove the vehicle, and took us back to our residence. I knew that I could rely on him. Like the radiant moon in an otherwise artificial city, Mr. Singh is refreshingly different. He works day and night. He is always moving. Yet, his struggle is his poetry; he is stable and calm; he has not learned this lesson from Shri Shri Ravishankar's art of living classes; he is naturally inclined to it. His mobile was constantly ringing. His customers were enquiring. He received all these calls with absolute patience, and simultaneously directed his colleagues to reach appropriate places and meet the demands of his customers. No restlessness. No irritation. No hard bargaining. He was only fulfilling his *swadharma* with absolute calm. Indeed, inside his taxi we were peaceful, even though the city outside was terribly speedy, restless and tension-ridden. Meanwhile, we reached our residence. I wanted to give him some extra money. 'No Sir, I can't take it'—his soft words reminded me once again that the moon would continue to radiate, and its beauty would prevail, even if the mega city breeds greed, violence and neurotic restlessness.

2

Petals of the Same Flower

We—friends, relatives and students—often meet together at our residence, choose a theme for discussion, contemplate and communicate. Far from being academic, it is intimate, informal and experiential. No wonder, music and poetry become an integral component of our gathering.

The other day when we began to reflect on 'nature as illumination' there was extraordinary enthusiasm. We all live in the city. Possibly the man-made world we confront all around—vehicles, flyovers, technological gadgets, market complexes—makes us tired, even when some of us may feel that there is no escape from this logic of technoscience. Moreover, from morning to evening—we allow ourselves to be thoroughly

programmed; with our computers, mobile phones, office cubicles and all sorts of packaged stuff we are fast becoming denaturalised. It is, therefore, not surprising that there are moments when, despite all sorts of pragmatism implicit in urban living—wealth/career/education/medical facilities, we see its hollowness, and nature—its rhythm and abundance, its beauty and vastness—begins to give us a call. Nature awakens us, makes us realise the spirit of our connectedness, and our location in the larger cosmos. Not surprisingly, Tagore and Wordsworth, mountains and rivers, distant stars and deep forests, spirituality and ecology—all grand ideals were discussed. And, of course, there was music: music as prayer. It softened our hearts. The entire gathering acquired a new meaning. We all felt that it was something wonderful.

Why was it wonderful? Because amongst us there were excellent orators; the words they uttered came out of intense feeling. Anita spoke of the Himalayas—its vastness and wonder, and the feeling of reverence it generates. For her, it was like a merger of our finite existence with the infinite. Irfan recalled the trajectory of Persian and Urdu literature, and saw nature as a grand

metaphor. Anirban—a physicist—spoke of science, and its perception of nature. Not solely that. Amongst us there were sensitive poets; their recitation elevated our consciousness. And, of course, there were gifted singers. They were active. They were visible. They took us to a higher domain of ecstasy.

However, many things were happening in silence. My sister-in-law—a simple woman from a small town in West Bengal—was not one of those who could speak on Tagore and Whitman. She was not at ease with English and Hindi—the two languages in which most of us were communicating. Yet, it would be wrong to regard her as a passive observer. She was doing her own *swadharma*; she was speaking her own language. With absolute perseverance, patience and artistic zeal she worked in the kitchen, and prepared delicious desserts for all of us. Food became divine; food became an offering, a service, a mode of communion: no less rhythmic than a piece of poetry or music. An Indian mother in the kitchen taking care of everybody—am I romanticising a role that feminists despise? Or, am I realising that life has many colours, and with her maternal instinct for serving/nurturing she is like a huge tree providing

us continual shelter without any expectation?

There was a young man. Absolutely quiet. He did not utter a single word. He listened with great care and reverence. It was indeed an extraordinary contribution. Listening, I felt, is important. Without listening there is no understanding, and without understanding there is no communication. With listening evolves humility and humbleness—the way the tree listens to those who surrender. He was a great listener. His eyes and ears vibrated as poetry and music illuminated the room. Silence is inspiration. Listening is courageous. Humility is strength.

What about Shivkumar—a gardener who lives in the world of plants and flowers? Our university vocabulary, we know, has already defined him, and put him into a box for a rigorous analytical tool. Historians regard him as a subaltern; political sociologists define him as a Dalit; and economists call him a migrant labour. Does he have the time or luxury for appreciating nature? Does poetry exist for him? Or, is he only struggling and somehow surviving? Does the subaltern speak? That morning Shivkumar came with a leaf on which he had put soft/beautiful flowers. A gift for all of us. It was a wonderful morning filled with Shivkumar's radiant smile, his gentle offering. And I learned a great lesson. One is not just a Dalit or a migrant labour; one is also divine; and poetry transcends the professional poet because it has its place in everybody's heart. Shivkumar is nature: simple and innocent without any armour.

The programme continued. There was a photograph hanging on the wall—a woman smiling and looking at us. Everything around us was getting enchanted. The photograph had ceased to remain merely a photograph. She was continually radiating rays, and illuminating us. Her temporal/physical body has already been dissolved into nature. Her light has become free from finitude; it is now all-pervading. And this light enabled us to overcome the darkness of our minds. We came closer to nature, and felt its enchantment.

As I contemplate, I realise that in a great act of creation there is no winner, no loser; there is no narcissistic hero, no alienated spectator. Everyone has a rhythm: a dignified space. My sister-in-law, the young listener, Shivkumar, and the light of that remarkable woman— were all petals of the same flower we nurtured that evening.

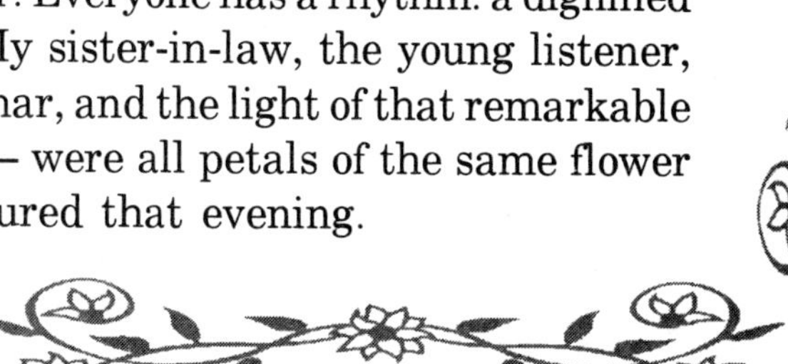

3

The Mystery of Fog and My Engagement with Laxmi

It was yet another cold evening in January. There was intense fog with its all-pervasive mystery and poetry. I thought I should come out, and merge myself into this wonderful play of nature. I started walking, and then this very rhythm of my evening walk took me to the residence of Maha and Rakesh. I found Maha deeply engrossed with her children—telling them stories, looking at their books and drawings, and helping them in their homework. It was indeed a sign that immediately entered my being; I began to see Laxmi: benevolent and beautiful, a goddess protecting and nurturing the home, and giving warmth to the children who have just begun to see the world. Amidst the fog and shivering cold outside, the home inside was full of warmth. Nature, motherhood, divinity—there was a music of symmetry, and I could listen to its rhythm and harmony.

And then what a coincidence! Maha presented me with a book on goddess Laxmi: a book she had written as a historian. In that act of giving I felt the merger of historiography and maternity, academic reflections and deep experiences. I realised once again that the dualities around us—reason vs. faith, intellect vs. intuition, private vs. public, motherhood vs. personal autonomy—were utterly superficial and false. I felt the joy of integration.

The gift was with me; it was time to leave, and return to my residence. Yes, it was cold outside, and the fog got more and more intensified. I started walking again. Everything around me began to look pure and sacred.

Are our moments of joy being accompanied with a deep sense of pain? Otherwise, why was it that at that very moment I began to feel about Kamalika—my friend's daughter? She is brilliant, gentle and compassionate. The other day she came to see me. With intense pain she spoke of a

dream crumbling, a possibility getting ruined, a leaf falling from the tree. Not everything was fine in her relationship with her husband. But then, there was a time when both of them dreamed and promised that their relationship would celebrate a new world filled with trust, reciprocity, and above all, human solidarity as a movement, a possibility and an aspiration. There was something fundamentally wrong in the relationship; her tears made me feel that her husband was deviating. In fact, Kamalika made me recall yet another story of a student of mine. Bidushi—a compassionate woman whose heart had always felt for the wretched of the earth—used to tell me many stories of Jesus Christ, and once gifted me her favourite book: the Bible. One day she got married, and I saw light in her face. But then, everything was over after four years. Once again the same story: the husband deviating. No home, no warmth, no Laxmi.

What then is the real? Maha as Laxmi, or Kamalika and Bidushi as broken, disenchanted, fractured souls? I know that there is a world around me that shows how promises are being broken, relationships crumble, egotism enters, and an environment

filled with fear, mistrust and violence postpones the arrival of Laxmi; I know that there are theoreticians and social activists who argue that the idea of Laxmi is a myth; the name of masculinity is violence; the home has no warmth; it is an oppressive site in which women are subjugated and humiliated; and above all, the emancipation of women ought to mean the ability to become independent of the trap of male-centric relationships and associated mythologies.

Yet, all these arguments which are often being heard at the university did not fully convince me. Beneath statistical generalisation and an overarching theoretical framework, I told myself, lie human possibilities. After all, I saw Maha: her fulfillment; and Rakesh—her husband—was no patriarch; with his contented self he was in a mood of complete surrender before the all-pervading Laxmi. And the tears of Kamalika and Bidushi made me feel that even in their pain there was an affirmation: the desirability of a home filled with the warmth of maternity, a home where Laxmi radiates.

I reached our residence. I looked at the wall. And she came out from the photograph. No longer a still photograph hanging on the wall, but unbounded energy—a living presence. I started receiving the rays. I got the message. I listened to her voice:

Laxmi is real; Laxmi is our light, our inspiration; Laxmi means masculinity getting redefined, and surrendering before the grace of maternity; Laxmi means femininity fulfilling itself in becoming a banyan tree protecting, nurturing and providing shelter, beauty and prosperity to all of us; Laxmi means a life-sustaining bridge between the home and the world.

She touched my forehead, and my entire association with her became alive once again. I got my answer.

Let Maha continue to radiate the home, and illumine the world. Let the fire (which is burning Kamalika and Bidushi) further strengthen them; let their tears soften their tormented souls, and make them realise that they are indeed Laxmi, and they are destined to create a much larger home where pain would be transformed into a blessing, and flowers would bloom even in a desert.

Nandita—my niece—is getting married. How do I bless her? I close my eyes, and they all appear before me—Maha, Kamalika, Bidushi, and above all, our eternal light radiating from the photograph. I come near the telephone, dial her number, and convey my message: Let marriage be a new birth: a moment of realisation that Laxmi is not far away. Open your heart, and invite her to the depths of your being.

4

The Moment of Meditation

'I have thought of a new story, a new plot'—the voice came from the intensive care unit of a super-speciality hospital in New Delhi. He was eighty years old, and his lungs were severely infected. With extraordinarily sophisticated medical gadgets and the supervision of specialised physicians and nurses, he breathed with immense difficulty. Yet, he was fully alert, and his mind was filled with a high degree of creativity. I got only five minutes to meet him. He looked at me, spoke about the new story he would love to write. And he reflected on man's spiritual quest.

My time was over. I came back. The hospital and its ambience; the doctors, their

sophistication and their terribly mystified/ specialised vocabulary; the spectacle of medical gadgets and all sorts of parameters emerging out of innumerable tests conducted on his body; and the anxiety-ridden faces of close relatives and their whispers—I was confronting the reality around me. However, amidst this gloomy environment, there was light, a flash of truth that touched my entire being. Who was that man the doctors were experimenting on? For them, he must have a name, an identity, a class; he must have a medical insurance and so on and so forth. But most importantly, for them, he was a body—a diseased body, an old body with all sorts of negative parameters; and their professional pride must manifest itself in the mighty war against the possible onslaught of death. They know his blood and urine culture, MRI and X-Ray reports; their discourse is centred on these parameters. But did they understand the deeper significance of the fact that he was imagining a story even when they were perpetually monitoring him in the intensive care unit? Was he only a diseased body? Or, even at the time of the disintegration of the body, did he continue to remain an artist, a mystic, a seeker?

It is quite unlikely that medical science would be interested in these questions. Because it is reductionist, mechanistic and fragmented. It prevails only in the domain of the physical. One is merely a body. And the decay, dissolution or death of the body, it is thought, is the end of the story: a sad and pathetic end. 'We are our bodies'—modern medical science as an offspring of the physical/techno-instrumental worldview further reinforces this belief. So what is left if the body is in crisis, if it begins to decay, and moves towards its final dissolution? It generates fear—fear of losing everything we have. Modern medical science does not succeed in overcoming it. Instead, it is being reinforced through utterly asymmetrical relations that prevail between the doctor and the patient, through a process in which doctors are transformed into sophisticated technicians who have no understanding of a being except its bodily/physical parameters, and above all through the very rationale of the medical gaze that continually suspects and stigmatises the diseased body.

Yet, the old man was visualising, imagining and creating. He could not be merely a body. Beneath the form called the physical body, my intuition awakened me,

lies the fountain of energy. It is this energy that composes music, writes poetry, visualises stories. It is this energy that no MRI scan can capture, no crematorium can burn. It is this energy that shows its spark even in the ICU of a hospital.

An awakening of this kind releases my mind. Death, I tell myself, is the death of the form: the finite/temporal/physical body. It is inevitable. All that is temporal has to disappear. And hence there ought to be an art of dying. How pathetic it is that our hospitals do not have any understanding or appreciation of this art. Instead, they make death ugly, regard it as an enemy, generate fear, and deprive it of its poetry by declaring war against it. Its cumulative manifestation is the ventilation machine. Death becomes lonely and technologically-mediated. True, the body needs to be healed. But its disintegration is also normal and inevitable. When the time comes, accept it with grace and gratitude, with poetry, music and prayer. It is like acknowledging the temporality of our physical existence, and realising that with death the energy does not disappear, it only releases itself from the form that one often confuses with one's primary identity. The moment one equates oneself with the

body, the egotistic identity evolves. 'I am my body. I am all that is apparent, physical and visible'. And with this ego emerges fear. However, when there is awakening, when one realises 'I am not my body. I am only a manifestation of the all-pervading energy'—the fear begins to disappear.

I came back from the hospital, and yet another incident came to my mind. She was just eighteen years old. Her mother was dying; all friends, well-wishers and relatives were mourning and feeling restless. 'She is not my mother. My mother is not that diseased body the All India Institute of Medical Sciences is about to declare dead. She is beautiful. She is my deeply intimate experience; she is the ultimate energy'. Possibly at that intense moment in the private ward of the AIIMS she felt: 'My mother—a source of unbounded love and energy—cannot die. It is only a form—a body—that is disintegrating'.

The hospital issued the death certificate—a symbol of modern order, its urge to register, document and classify everything. But who died? The fact is that the girl sees her mother as immortal energy manifesting itself in the full moon radiating the snow peaks in the

Himalayas; and the old man is imagining a creative story while suffering in the hospital. That is my truth, my inspiration, my moment of meditation.

5

As Music Unfolds Itself…

I often tell myself that understanding is a process of becoming; it is not a finished product. Every moment, I have begun to realise, has its surprise and wonder; and life keeps unfolding itself. Take, for instance, music. I don't know whether I can cognise its technicalities and grammar. However, every day, for me, music reveals itself, touches my entire being, and I keep understanding and rediscovering it.

I come to the terrace; it is time for the sun to set on the western horizon. I begin to witness this wonderful moment—its grace, depth and melancholy. The clouds are changing their colour, tears begin to fall from my eyes, and, believe it, I find my Bismilla

Khan in the distant sky; he is playing the *shehnai* for me. Never before did I understand Bismilla Khan in this fashion. Possibly music, I tell myself, is transcendental. It is profound and sacred. It reveals itself when the ordinary becomes extraordinary, and the mind is ready to receive and surrender. Only at those moments do we feel music, not just hear it. As I look at the vast sea, or as the snow peaks in the Himalayas give me a call, I feel that our finest music is our sacred prayer. With Bhimsen Joshi and Ravishankar I too begin to pray...

When I was a child I did not understand much about Rabindra Sangeet. I do not know whether I understand it now. But as I experience life, its diverse colours, its deep moments of joy and pain, Rabindra Sangeet comes to me as my longing, as the finest truth of my existence. There are moments when books, theories, statistics, empirical facts, university seminars, lectures—nothing gives me what I long for; and then I listen to a song, and like a flash of truth it illumines me. Imagine my early morning. I rise with a feeling that I would never find her in her physical/embodied existence; yet, everything that I do is in continuation of her spirit and

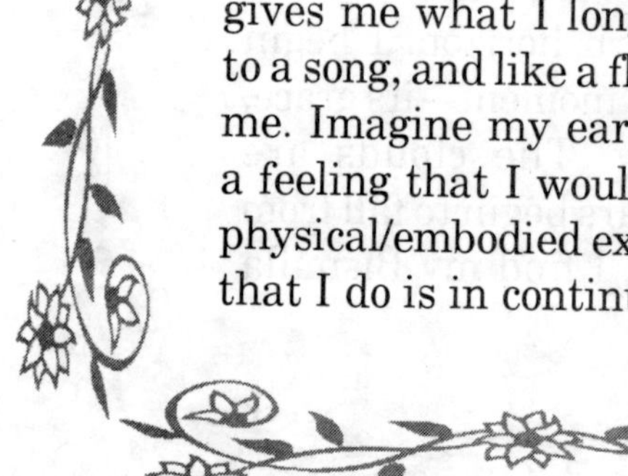

energy. I begin my day with a pledge: her rhythm is my rhythm, her energy is my energy, and we must carry on. I switch on the music system, and Debabrata Biswas—an extraordinary singer who gave a new touch to the tradition of Rabindra Sangeet—takes me to my truth:

Let the fire of your rhythm spread all over...

I do realise that a poet, like a mystic, is a visionary; a poet sees the deeper layers of existence, and when poetry becomes music, my truth becomes revealed truth—deep, intimate, experiential.

So what is real? Reality is not just what is explicitly visible, what statisticians measure, tabulate and quantify. That is merely surface reality. Beneath it lies the deeper reality which reveals itself when the mind is without any armour, when the heart opens up, and when reason becomes free from all sorts of cunningness and instrumentality. Only then does music touch us, and take us to the depths of our existence. Imagine yet another morning—its busy schedule. I get up, clean the rooms, store drinking water, prepare tea and breakfast, and ask my daughter to get herself ready for the college. Meanwhile, someone knocks at

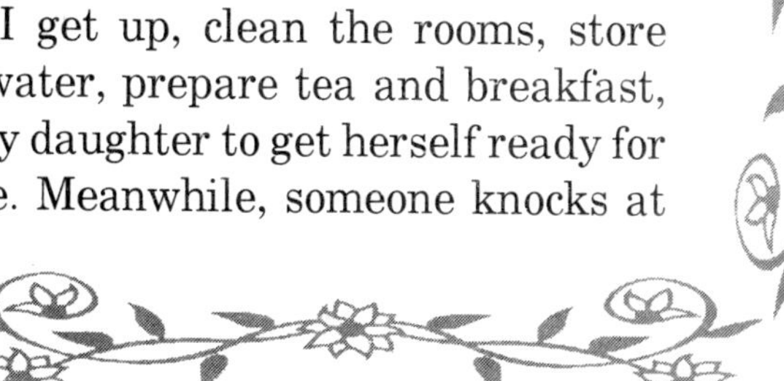

the door, and persuades me to buy fruits and vegetables from him. Not solely that. The LPG cylinder is already exhausted; a new one has to be fitted. And the clock—its disciplinary time—is reminding me that I have to move fast, it is about to be 9 o'clock—the time for all of us to start for our work places. Yet, I acquire the courage, take a break, come to the terrace, and discover with absolute joy a beautiful rose blooming; and all our plants and flowers are eager to embrace me. I come back, switch on my music system, and Kabir Suman—a contemporary Bengali singer from Kolkata—begins to sing:

Everything is hectic at 9 o'clock. Yet, it is at 9 the flower blooms on the terrace.

I tell my daughter: 'Listen to the song. Feel it. Reality is not just our hectic schedule and time management. Reality is also the flower blooming, the plants dancing. The poet sees it. The singer feels it. They have the eyes. We too must keep our eyes open.'

Indeed, life is this mystery. Music exists amidst the prose of practical economics. The sun is still rising, the moon is radiating, the butterfly is communicating with the flower, the leaves are dancing, and Rabindranath Tagore perpetually reminds us of our

'surplus': our ability to see beyond mere utility, and fulfil ourselves in the abundance of nature and the cosmos.

This is my way of finding my music. Every moment I rediscover it. I cannot sing. But I know that Gita Dutt and Hemant Mukherjee often enter my inner world, and their voice becomes my voice. Through the window of a running train I see a village man, and imagine Mukesh singing for Raj Kapoor. In a lonely railway station in Andhra Pradesh a tea vendor's transistor makes me feel the rhythm of India; a Lata Mangeshkar classic thrills me, and my journey becomes a distinctively Indian journey. How wonderful it is to find my music all the time. I find it when I am happy, when I am tired. I find it when I am working; I find it when I am in deep pain. I find it in the vibrancy of the city, in the silence of the mountain, in the university lecture hall, in the eyes of a cancer patient.

Let this music accompany me till my last breath. Let music soften me, expand my horizon, remove my meanness and arrogance, give me the strength to grasp the Upanishadic prayer, merge into the Buddha's eyes, and understand the sermon on the

mount. Let me close my eyes, and see John Lennon coming down from the clouds with his guitar, singing, and inspiring me to 'imagine' and celebrate my utopia.

6

Gift

The other day an old student of mine came to see me, and she gave me a wonderful gift. It was her own creation—an illuminating drawing of Sri Aurobindo. Why did she give me a gift of this kind? I remember I taught their batch a course on education, and I did invite them to the pedagogic ideals of the great saint-philosopher. Possibly Sri Aurobindo's ideas shaped her consciousness, influenced her being. And through the gift, I thought, she sought to share the experience of her communion with Sri Aurobindo. That she chose me for this purpose was enough for me. It softened me, made me humble; and most importantly, I began to realise the worth of a gift. A gift is a tangible expression of a deep experience—a shared relationship, a memory, a willingness to communicate.

It is sad that we allow the market—its calculative logic, its exchange economy, its strategy—to devalue this essential beauty of a gift. Once it is marketised, it loses its poetry. It gets measured, quantified, and transformed into a weapon, an assertive declaration of one's might. As a 'prestige issue' it begins to carry a price tag. The gift you give, and the gift you receive—everything becomes a medium through which the language of social status and economic power gets communicated. Not solely that. Another manifestation of this instrumentality is that gifts become increasingly utilitarian. Look at, for instance, the way these days many of us give gifts to a bride/bridegroom at the time of their wedding. Quite often we keep hard cash or a cheque in an envelope, and present it as a gift. The logic is that money can buy everything. Yes, it does; but it cannot replace our poetry and aesthetics—our own smell and touch. Marriage is sacred, and with a gift as our own creation—our genuine engagement and feeling—we could have further enriched its sanctity. It is sad that the lure of the market makes us forget that a gift is a gift because it takes us to a finer/higher/subtle domain: a domain beyond calculation and utility, a domain that radiates only love and concern.

At a time when everybody speaks of the irresistible march of the market, and when all sorts of shops are available containing attractive packages as ready-made gifts, my student gave me a pleasant surprise. It assured me once again that no situation is so powerful that it can annihilate the human imagination. Poetry is alive; aesthetics does exist; and the profound simplicity of a gift is still a possibility. I am indeed fortunate. Time and again, I have received such wonderful gifts: a book of Sufi poetry, a piece of Tibetan music, a shawl a student purchased from her first salary, a basket of Assam tea yet another student sent from Tejpur, and above all, their healing touch, their assurance, a telephone call from a distant land—'Last night I have seen you in my dream. How are you doing, Sir?'

Who knows? Possibly we too have to orient ourselves, and become receptive. Only then do we get gifts—from people, from nature, from the rhythm of life itself. Indeed, gifts are in abundance. We have to keep our eyes open, expand our heart; we have to be humble to receive these gifts.

My wife was cooking. And she called me into the kitchen. I followed her instructions. She asked me to look through the window.

I did, and with absolute reverence I saw the master artist's play: the hide and seek of the full moon and the dark clouds. What else could be a more marvellous gift?

All these gifts have taught me a fundamental lesson. There is a world beyond the calculative logic of hard economics. It is only in such a world that we fulfil ourselves, realise our innate possibilities; we become rhythmic, musical and poetic. It is only in such a world that the act of sharing becomes an uninterrupted flow, and even our acute moments of pain and suffering are transformed into a possibility of redemption.

The gifts I receive make me pray. My prayer emanates from the depths of my being, my longing—Make me capable, purify me, elevate me so that I too can give gifts, and let this giving be an act of humble offering. Neither money, nor power. Only love; love mingled with aesthetics: Blake's songs of innocence, Ramakrishna's ecstasy, Tagore's poetic spirituality, Jalaluddin Rumi's wisdom—'Each drop of blood of mine is saying to Thy dust, I am the colour of Your love, companion of Your affection',and above all, the art of dying—death of ego, death of pride, death of possessions. Pure. Empty. Subtle.

7

Reading as Enchantment

What is the meaning of reading? As I have always felt, it ought to be an illuminating dialogue, a celebration of ideas, an inspiring moment leading to the flash of truth: simple, non-pretentious, authentic. I teach in a university, live in the world of ideas, and, as outsiders might think, reading is my profession. Yes, we read a lot; but quite often the entire experience gets burdened and monotonous. The other day I was reading the draft of a Ph.D. thesis. Technically, the thesis was good; it contained everything that the culture of higher academics values: complex concepts, review of literature, elaborate footnotes, impressive bibliography, empirical findings and theoretical frameworks. But then, something happened to me. The technical perfection of the thesis, for me, became unbearable. I found only words and clever intellect. No light. No illumination. No wisdom.

This boredom made me reflect. Why is it that our knowledge, far from liberating us, becomes a heavy burden? Is it because our academic enterprise has killed our innocence—the eyes to discover the world once again with all its wonder and mystery? Is it because we have become merely trained professionals who have learned to equate complexity with wisdom, technical vocabulary with truth, understanding with footnotes, and demonstration of knowledge with clarity? 'Excellence' proliferates, wisdom declines, books and theses become oppressive, reading/writing becomes merely a professional duty—a show, a demonstration.

With this reflection I became restless. I stopped reading the Ph.D. thesis. I felt like having a breath of fresh air. I became eager to look at the vast sky. I was filled with an urge to surrender before the trees, and move with the butterflies. I was trying to unlearn; I was searching my poetry, my innocence. Who would assure me that the rhythm of

life—its deeper meaning—is simple and musical, and it prevails despite the bombardment of knowledge, despite all those thick theses and books written by professional philosophers and social scientists?

Possibly my prayer worked. The same evening I happened to meet two people. A student of mine came to see me with his father. He looked extraordinarily simple, not like those knowledgeable men and women I come across every day. His thin body was rhythmic; his distinctively rural gesture was his beauty; there was a glow in his face. As the conversation followed, he began to speak of his mission. He moves from one village to another, teaches yoga, inspires people to lead a life closer to nature. No big words. No quotation from the scriptures. His words were like the stream of a fountain. 'There is no past, there is no future. The present—and the present alone— exists'. His words moved me; I began to listen. He spoke of life and death, pain and suffering, freedom and redemption. Words became musical. He began to sing a series of *bhajans* indicating our prayer and longing for a merger with the infinite. He left, and as a living book he healed my wounded consciousness.

And then came our gardener. Always cheerful, full of love and compassion. He nurtures flowers, lives in the world of flowers, speaks of flowers. 'In my village people say that a good man is like a flower because a flower is inherently simple and beautiful'—he told me. And I was overwhelmed. Within a second our gardener became a William Blake, a Wordsworth, a Tagore. I experienced poetry—neither in a thesis, nor in a seminar on cultural studies, but in his folk wisdom.

These two people renewed my energy. I acquired the courage to return to the university library. I was determined to get the kind of books that would give me back what I was expecting from my engagement with reading: a softening touch, a deep empathy, an experience of merger. And I found Baudelaire—the great French poet: the collection of his poetry. I am aware of his pain and misery. His extraordinary sensibility touched me, and I read again and again:

I hate you ocean! Your storm and stir
My spirit knows; the bitter gaiety
Of man defeated, full of insults and tears
I hear you in the vast laughter of the sea.

However, life is not just pain and misery: the anguish of a gifted poet like Baudelaire. Life is also an aspiration, a quest, an opportunity to experience joy and wonder in the entire creation. I found yet another book: this time a book by Herman Hesse containing his essays on life and art. What an extraordinary writer filled with penetrating insight, wisdom and sacred longing! I began to read his essays, and every moment of reading became a purifying process. From Baudelaire's pain Hesse took me to a realm of possibilities:

As long as men are capable, in the midst of the distresses and dangers of their lives, of rejoicing in such things as these: the play of colours in nature or in painting, an appeal in the voices of storm and sea, or in man-made music, as long as beneath its surface interests and necessities the world can be seen or felt as a whole, consisting as it does of interrelationships, from the curve of a young cat's neck to the vibrations of a sonata, from the touching eyes of a dog to the tragedy of a poet, an interconnection of a thousandfold riches of relationship, correspondences, analogies, and reflections, out of whose eternally flowing language their hearers derive joy and wisdom, entertainment and emotion—

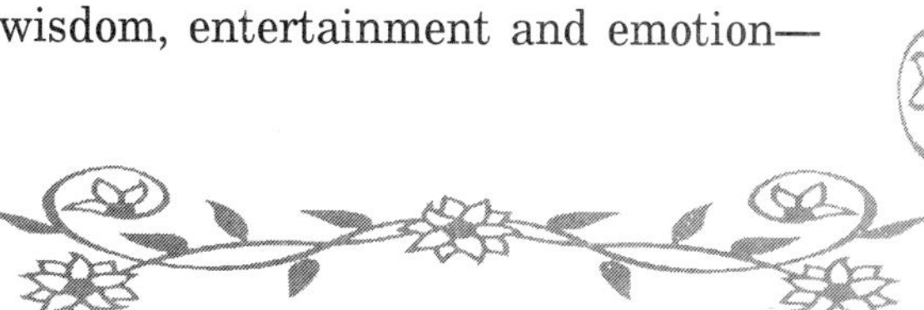

just so long will man again and again triumph over his ambiguities and be able to ascribe meaning to his existence.

The clouds have disappeared, and there is clarity in my consciousness. The demands of my inner self, I have begun to feel, cannot be fulfilled by my professional identity. Beyond the parameters of the university, its knowledge production, its theses and publications lies another world inhabited by the likes of Whitman and Dostoyevsky, Upanishadic sages and Sufi saints. No matter how trained I am in the grammar of the university—I tell myself—I should not forget to have a glimpse of that world. Only then is it possible to transform my act of reading into an offering, a prayer.

8

My Philosophers at Khirsu

Mountains continue to give me a call, and this irresistible appeal often takes me to the remote corners of Uttarakhand. Every visit widens my horizon, and makes me confront a series of existential questions. I come back to the metropolis, join my work sphere; but I do realise that a deep change has begun to take place in my consciousness.

This time I was at Khirsu in the Pauri Garhwal region of Uttarakhand. The vastness of the mountains, the snow peaks reminding me of the majestic presence of Gangotri and Kedarnath, the dense forests, and above all, the all-pervading silence—Khirsu, for me, became a wonder, a metaphysical riddle, a site of learning no university could cope with. In fact, this time

I could not escape engaging myself with the notion of fear: why is it that the psychology of fear surrounds our existence, our every act, our every movement?

Well, the beauty of Khirsu, its purity, its smell filled my mind with joy; yet, there was intense fear. Possibly I feared its remoteness, its difficult terrain, its everyday struggle. My metropolitan existence has made me idealise the mythology of a global village as it conquers time and space; its cars, aeroplanes, mobiles, e-mails diminish the slightest notion of remoteness. Can I then live in a place where there is no guarantee of electricity, where mobile connections are perpetually disturbed, where comfortable transportation is a difficult proposition? I saw small children climbing the mountain, walking five/six kilometres to reach their school. I got frightened. Because I am used to the reality of comfortable school buses, or the over-protected parents driving their big cars, and dropping their children at the school premises. I saw women assembling near the hand pump early in the morning to store drinking water. The scarcity of water frightened me. And where is the ambulance, the hospital, the emergency service, the medicine shop? What do these people do if

they fall sick, if late at night someone has a cardiac attack? My body began to tremble. There was fear—and only fear—that enveloped my consciousness.

The music of the birds awakened me. I woke up quite early. I came out. The mountains in the early morning, the freshness all around, the feeling of a new beginning softened my being. Everything acquired a new meaning. The children were climbing the mountain, and coming to the school. They looked extraordinarily beautiful. They were not tired, exhausted. No nausea. No boredom. Instead, they were full of zeal and energy. Musical and rhythmic. 'Where is your village?' I asked them. They smiled gently, and showed me a village in the deep valley. 'Every day you come to your school from such a distance. It must be quite tiring,' I further asked. They laughed and ran away. My question, I imagine , made no sense to them. They live. They do not theorise. They do not complain. Who am I to impose my fears on them?

The women had already begun to store water. There was no tension, no conflict over the scarcity of water. They waited patiently. Indeed, they can wait; I cannot. And that is

why, whenever there are adverse situations, I get afraid; but they accept life as it is. As I was contemplating, I saw an old man and his wife coming out of the valley, and climbing the mountain—slowly, but steadily. I waited for them. I introduced myself. They smiled, they communicated without inhibition. Their wrinkled faces revealed wisdom, patience and endurance. They looked above, and with absolute humility spoke: 'Everything depends on Him. There was not adequate rain for the last two years. If God gives us rain this time, everything will be in order.' Their words gave me a vantage point. Life, for them, is difficult. However, they do not fear it. They accept it. Their prayer, their humility, their patience give them the strength to carry on.

But then, I ask myself why people like us are always afraid, always complaining, always living with a sense of loss. Is it because the entire logic of modernity is centred on fear? Well, it is always possible to argue that modernity is taking us to a world in which there would not be any reason to fear; science would provide clarity, and overcome the fear of the unknown; technology would give us the power to control the unpredictability of nature; compulsory

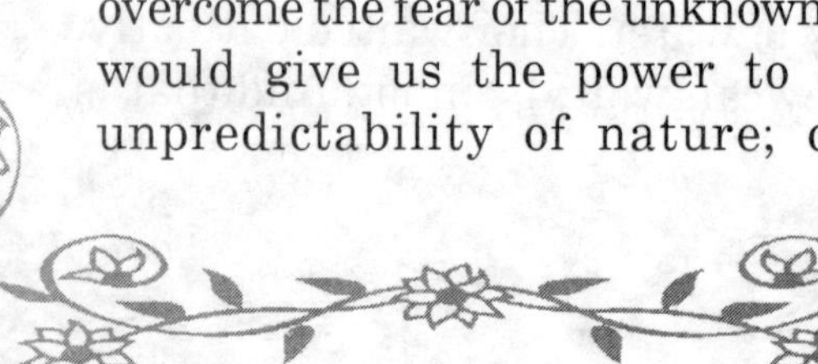

schooling would give us knowledge and wisdom; and modern medicine would eventually conquer death! Yet, paradoxically, the more we seek to conquer fear, the more fearful we become. As a matter of fact, we fall into the trap of fear. No wonder, even when we send our children to what we regard as 'good' schools, we continue to fear whether they would acquire the required grades to become 'successful' in life; even when we buy our apartments in the best possible locality, we fear what would happen if it is not easily connected to the metro railway, and all our insurance companies keep cultivating the fear of the impending danger: insure, or be insecure! Cars, apartments, jobs—in fact, life itself—is in perpetual danger, and one insurance leads to another. The more we get the more afraid we become of losing it. Modernity has not eradicated the psychology of fear; it has used and intensified it because, instead of understanding the roots of fear, it seeks to conquer it through a series of external techniques, and the more it tries to do so the more it falls into its trap.

Khirsu taught me a refreshingly different lesson. Fear cannot be overcome through the rationale of fear. The root of fear is

essentially our refusal to accept life as it is: its peaks and valleys, its light and darkness, its activity and passivity, its vibrancy and silence. And hence as Khirsu taught me, it is important to accept life as it is even when it appears to be hard and difficult. The children climbing the mountain, the women storing drinking water, and the old couple looking at the sky—they emerged as my philosophers, my tutors. I began to see the worth of humility, patience, endurance and some sort of detachment. Why fear? It is the 'ego' that fears. Khirsu taught me the futility of ego, its absurdity.

It was late at night. I came out of my room, looked at the clear sky full of stars. The mountains were getting mystified. I began to feel the intensity of the all-pervading silence, the depth of darkness. I realised death: death of ego, death of neurotic restlessness, death of what modern social scientists regard as 'agency'. And at that moment I became free from fear.

Death, I tell myself, is not different from life: the way, as the mountains reveal, the peaks make no sense without the valleys, or the freshness of the morning is inseparable from the depth of the dark silent night.

9

God or No God

Possibly it was one of her contemplative moments. After the terrible summer there was a spell of rain; and then the smell of the earth must have taken her to her most intimate zone—the memories of her mother. Why did her mother leave her physical body so early? Why did she get deprived of her mother's earthly presence—the warmth of her body, the compassion in her eyes, her words, her vibrations, her cooking, her smile? Did God have the courage to appear before her, and explain why He took away her mother? Even if there is a meaning in her death, why can't God come and explain? She kept asking me.

I did not want to intervene. I began to listen. God, if there is any, as she told me,

must be terribly insecure; He does not want to share His secrets; He does not have the courage to reveal His design before us. He wants to keep us perpetually helpless. And this helplessness, it seems, leads to the fear of God. Anything that is based on fear cannot be authentic. That is why God, for her, has ceased to exist. Even if there are 'revealed' souls who have seen God, it does not satisfy her. Why should God be so choosy? Why should He appear only before the select ones? Isn't it a sort of elitism, a demonstration of inequality? 'Maybe I am too young. Maybe I do not understand the spiritual path. However, the fact is that I am not particularly convinced of those who speak of God.' She continued to express herself.

Am I capable of providing a convincing answer to her queries? Possibly I cannot. Because like her, I too am asking, wondering and reflecting. For me too, there is a huge domain of uncertainty. I do not know whether God as a supremely powerful being exists, and governs our destinies. I too see inexplicable tragedies, disorder, utter chaos and meaninglessness in innumerable things happening around us.

Yet, there is a feeling which, at times, surrounds my existence, and tends to restore

some sort of calmness even amidst tragedies and storms in life, and prepares me to evolve a meaning in all that we do. God or no God, I see the worth of this feeling, and wish to convey it to her. She lost her mother, and apparently there was no reason why it would happen, particularly at a time when she was growing up, and needed her at every moment of this wonderful journey. However, beneath her deeply personal tragedy lies a universal fact, and a constant awareness of this universality enables us to look at things—even our own tragedies—with a larger perspective. All that is created is temporal; it is impermanent; it is bound to disintegrate, disappear and die. It is happening all the time. Our empires, our status and fame, our loved ones, our bodies we are so closely attached to—nothing is permanent. True, something disappears earlier, something somewhat later. But eventually everything has to go. Like all forms and all expressions of temporality, the embodied existence of her mother did disappear. I do not intend to suggest that she should not grieve, she should not feel sad. After all, her tears are sacred, her loss is her realisation, and her mother is her epic. But then, an awakening, a realisation, an acceptance of this permanent

truth amidst all sorts of impermanence—the fact that we all come, meet in a tourist place, and then disappear forever—give us some clarity. Despite our terrible tragedies we gain maturity: the strength to see the events of life as an ongoing play in which actors and players keep changing all the time. This, I believe, is the beginning of compassion. Why should we be harsh to one another when we realise that we are here only for some time, and all of us are in the same boat?

With this realisation, it seems, comes a sense of detachment. What does detachment mean? Is it coldness, passivity, a sort of life-negation, a flight into an other-worldly endeavour? Yes, there is a possibility to misinterpret it as some sort of an escape. However, detachment, I would like to imagine, is essentially the strength to live life, to participate in this ongoing play with an awakening, with clarity and wisdom. I exist; but I ought to exist with a perpetual awareness that my body—my embodied self—which I often feel tempted to equate with my essence is transitory; death is inevitable, and, irrespective of modern medicine, it can come any time. I teach in a university, but I need to work with a realisation that what I teach is more important than my identity, my ego, my limited role as a teacher; and the university as an institution would continue to exist even if I resign, retire or die. I live with my loved ones; we eat, travel, shop, buy apartments, watch movies; but I must realise that even this network of relationships—the warmth of a home—is transitory; our partners will die, our children will grow up, start their new life-projects, and possibly settle down in distant places. What is full today would be empty tomorrow. Detachment emanates out of this acceptance. This does not mean that I enter a zone of eternal mourning, or I become cold, and lose a sense of humour and joy. Instead, detachment means that I am now prepared to accept life in its totality: creation and destruction, pleasure and pain. Good times do not make me excessively excited or arrogant; nor do bad times make me terribly pathetic.

Am I capable of living with detachment, even though I understand its worth, at least intellectually? My ego, it seems, is a hindrance; it is a source of attachment, and hence pride, fear and anxiety. Victory excites us, defeat saddens us. Our share markets, our sports carnivals, our political battles,

our mega hospitals—everything indicates how our egos operate, how attached we are to our wealth, our political ideologies, our loved ones, and our bodies. Yet, I cannot negate the importance of my quest for detachment. After all, detachment means the removal of the blockage caused by the armour called the ego. Detachment means the free flow of life-energy without any obstacle, without fear and anxiety. Detachment means calmness.

This is what I wish to share with that young girl. I do not know whether there is a God to protect this wisdom, and empower us with a sense of detachment. But God or no God, life itself is a great lesson, and this quest is sacred. I cannot console her. However, I can always pray, and hope that her sensitivity will eventually open up her eyes, make her receptive, give her calmness, and enable her to see and accept the reality in creation and destruction, in pleasure and pain, in light and darkness, in the activity of energy and in the silence of deep sleep, and above all, in the substance of forms and in the emptiness of the formless.

10

Evaporation of Identities

These days I often feel that my academic sociology is not always in tune with my deep aspiration, my intuition, my intimate experiential domain. Take, for instance, the notion of socio-cultural identity. It is a theme sociologists write about; my students write dissertations and theses on it; and, we all know, political activists—centrists, rightists, leftists alike—stimulate these identities. Not to situate man in the context of caste, gender, ethnicity, race, nationality—we are told—is naïve, a dangerous act devoid of a sense of history and political imagination. Sociologists tell us how identities are constituted, and how our acts, beliefs, practices, rituals and festivals cannot be understood without making sense of these identities that envelop us from all possible angles. You are a Brahmin, I am a Dalit; you are a Maharashtrian, I am a Bengali; you speak Tamil, I speak Kannada; and hence we are destined to be different and limited in our thoughts and aspirations. Universality is a myth, and the idea of shared aspirations is only a hegemonic device for homogenisation! No wonder, the politics that is based on differential identities—particularly if it has a subaltern touch, as we are told by academic experts, ought to be legitimate and desirable.

I have grown up in this academic culture. My training in formal sociology has taught me many lessons on caste, ethnicity and gender. I have also learned the language of political correctness my colleagues celebrate. Yet, there is a sense of discomfort. I feel uneasy. It becomes difficult for me to believe that our essence is primarily our social identities. There is something deeper, something universal that transcends all barriers, all limiting identities, all justifications for conflicting differences. I am not just in time and space. I also carry a transcendent self. I am beyond my caste, beyond the language I speak, beyond my dietary practices, beyond my rituals. I am beyond concepts and categories—limitless,

infinite and transcendent. Sociology theorises and possibly legitimates my bondage: how conditioned I am by my caste, gender, religion, nationality. However, I aspire to be free; I long for eternity.

And believe it, the intense moments of pain and suffering take us to the deeper domain of this truth. How often I recall one of the sites from where I have learned this great lesson: the private word at the All India Institute of Medical Sciences. My wife was suffering. As far as the doctors were concerned, her disease was fatal; there was no hope; there was only a soulless deliberation on the statistical probability of her survival. Yet amidst this darkness there was light. There was Uttara. From morning to evening—she would wait at the corridor: always willing to help in silence, and with absolute grace. Her blood was flowing through my wife's body. There was Deo arranging everything and organising blood donations for her. And there was Tariq moving around medicine shops and assuring me all the time. Vasudha, Ritu, Priyaranjan, Irfan, Jitha, Nabanipa, Pawan, Anirban, Sushanta, Piku, Papiya—the list is endless. What were their identities? According to demographers and sociologists, Uttara is a

Jain/Rajasthani girl; Tariq is a Bihari Muslim; Deo is a Maithili Brahmin; Ritu is from Lucknow; Priyaranjan belongs to Bihar; and Vasudha is a Kashmiri Brahnin. But then, all these identities—caste, gender, language and religion—evaporated at the AIIMS. There was a flash of truth; and I realised the fundamental oneness of being. My wife's suffering, our suffering became their suffering. Compassion, I realised, has no caste; sharing has no regional orientation. Uttara, Deo, Tariq, Ritu, Irfan—we all reached a domain that refuses to be defined through the discourse of limiting identities. My wife left her physical body; she became the eternal light. And that morning when the sun was rising after a terrible night, the corridor of the AIIMS was filled with vibrations. Despite pain, suffering and a sense of void and loss, I learned a great lesson: our capacities and potentials are infinite; we are not what we appear to be—just with a caste, a religion, a nationality; we are endowed with the 'surplus'.

It is sad that in formal academics these possibilities are seldom discussed. Its objectivity reaffirms the status quo. Its empiricism is static. Its categories are merely smart devices for tabulation and

classification. What it regards as knowledge has no vision because it does not have the eyes to see the possibilities we are gifted with. The politics it talks about is only superficially liberatarian because it falls into the trap of identities. It fails to appreciate that the differences we exhibit exist only at the surface level; beneath the apparently visible lies the realm of deeper unity—the way rivers lose their distinctive identities as they merge into the sea.

My discomfort with the formal knowledge system continues. I do not mind even if I am being regarded as a dreamy outsider. My dreams, I tell myself, are real because they have emanated from pain and suffering, from the most difficult moment in life. My dreams give me life-energy. My dreams make me discover eternity in Uttara, Deo and Tariq and all of them. With my dreams I realise that we truly rediscover ourselves only when we overcome the barriers of our socially constituted limiting identities, only when we become the all-pervading light: no caste, no language, no religion, no nationality; only deep silence, grace and eternal flow of love that transcends death.

11

On Football, Octopus Paul and My Enchantment

I love football. I love its team spirit, its constant vibrancy, its creative skills. Football reminds me of my childhood: the green field, and the magic of the match after school hours. Everything comes to my mind once again as the World Cup football presents itself as a spectacle. Yes, there is a fundamental difference between the kind of football I loved as a child and the highly competitive professional power football that gets marketised in a media-induced global society. It is no longer innocent. Instead, it is about the narcissism of nation-states, the production and consumption of stardom and associated images, and above all, the unimaginable flow of money. It ceases to be a simple game. It becomes a sort of war, a life and death issue, and a site for the Darwinian rationale of the survival of the fittest.

I watch this football, and am amazed at the way octopus Paul emerges as the most celebrated superstar, and begins to capture the attention of the entire world. Paul, we are told, predicts the outcome of the match, and every prediction is coming true. Neither Argentina nor Germany, neither Netherlands nor Uruguay can defy this prediction. The oracle power of octopus Paul surprises everyone. Television channels intensify the excitement further; rationalists refuse to be impressed; statisticians speak of probability factors; astrologers find a case; and innumerable fans of football keep wondering.

Paul makes me think. I begin to contemplate. From football to the philosophy of life—I undertake a journey. Possibly this fascination with the oracle power of Paul indicates the crisis implicit in the age of reason. Yes, we tend to believe that the cultivation of objective/scientific reason is a distinctive feature of our times. It explains. It demystifies. It demonstrates valid empirical proofs. It abhors intuitive insights,

prophetic visions and mystic longings. This reason is our pride; yet, paradoxically, this is also our discomfort. We should not forget that there is a longing for magic, for charisma, for the inexplicable. We often seek to overcome the iron cage of scientific reasoning, and taste what science cannot explain. We wait for the arrival of a mystic, the charismatic authority of a prophet. We wait for astrologers, or for miracles to happen. And in popular imagination even technoscience is being perceived as some sort of magic. No wonder, as it would appear, octopus Paul is a miracle—an exit from a disenchanted rational world, an opportunity to fulfil our need for the magical.

That is not all. There is something deeper. The fact that what Paul predicts becomes an issue reveals how insecure we are. One important reason for this insecurity is the fear of the unknown. This fear gets intensified when, as the World Cup reveals, there is terrible pressure to win, when victory is mythologised, and failure stigmatised, when there is no escape from the anxiety-ridden question: Who is going to win? What would happen tomorrow? Football is just a mirror. In fact, we live amidst this fear of the unknown. Things do not always happen

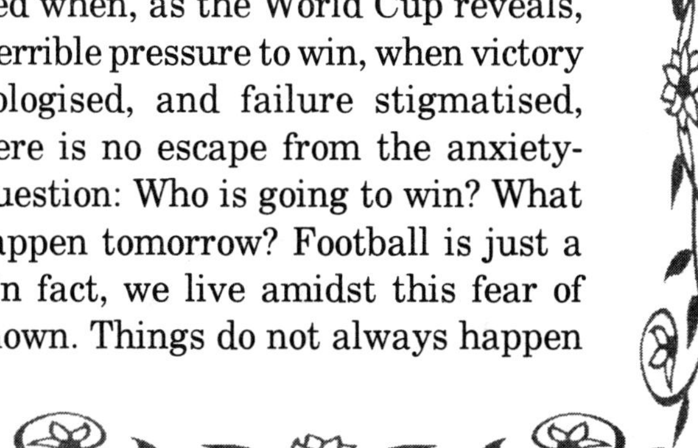

the way we want. The gap between our efforts and the fruits of our actions remains a puzzle. Inexplicable tragedies, accidents and even good fortunes surprise us, and there seems no answer to the question: Why? However, we want to know, and become certain. It is difficult to live with the fear of the unknown. We loathe uncertainty and chaos; we prefer certainty, order and predictability. Science seeks to do it, tries to bring about some kind of order and explanation in the world; weather experts assure us by predicting rains, storms and cyclones; and with innumerable tests and medical gadgets doctors try to predict the coming danger, even how long one would live; and above all, astrologers predict, and try to control our life- trajectories. Indeed, from science to astrology—there is an urge to predict the future, and control the unknown. At a time when the World Cup leads to excessive stress and tension, and all teams suffer from terrible anxiety, octopus Paul appears, and his series of correct predictions tend to satisfy the mind that is hungry for knowing the unknown.

I understand all this. But I do not wish to fall into the trap of what octopus Paul symbolises: the uncanny power of knowing

the unknown. Because this deprives life of its mystery, surprise and wonder. I would rather imagine myself living in a world in which only this very moment exists, and the future remains a mystery—an unknown entity. Let this mystery surround my existence; let me live without any fear of the unknown; let me face life as it comes: its rise and fall, its achievement and failure. I feel like praying: Give me a sense of humility—the willingness to accept that not everything can be known. Give me the strength to accept—with dignity and grace—whatever happens in life.

This acccptance, I would like to believe, is not passivity—my surrender before destiny. Even if there is something called destiny, I do not like an astrologer, or for that matter, octopus Paul to predict it. Instead, I would face it, live with it. Even if I am only an actor in a play for which the script has already been written, I would like to exist as a creative actor: one who makes efforts, does things with deep enthusiasm, and even attaches a new meaning to the script. My failures and tragedies should not make me depressed; instead, it should take me to the depths of human existence. My success, my achievement should not make

me mad, proud and arrogant; instead, it should make me humble. I should become complete and integrated, and realise that, be it success or failure, everything is in motion, and life is an ever flowing river. So why should I fear the unknown, and try to control it? Why should I feel so restless to know my fate? I would rather play—and play creatively—all sorts of roles in this cosmic drama. This is beyond science. This is beyond astrology. This is the poetry of life.

And what about enchantment? I do not need octopus Paul to enchant me. Football is my enchantment—the kind of football I loved as a child: football as just a play—without the burden of nationalism, without the seduction of stardom, without the lure of money, and hence without fear and anxiety. Football as pure bliss. As a child I experienced this bliss. As the World Cup euphoria gets disseminated through television channels, I come back to my sacred corner, close my eyes, and begin to experience the vast green field—the abundance of my childhood, and the smell of football in the monsoon. I invoke my inner soul, and whisper: Give me back my childhood, and allow me to sing my songs of innocence.

12

Some Rare Moments

The other day Raj Kumar—a student of mine—came to see me. He had cleared the M.Phil entrance test, and, as I felt, that was reason enough for him to feel exalted. But then, he gave me a surprise. He had already decided not to take admission; he would rather leave the university. Indeed, to begin with, I could not make sense of what he was saying. Raj Kumar is a bright student, his grades are fairly satisfactory, his intellect is quite noticeable, his political imagination is in tune with what a university like ours expects from its progressive students, and hence, as it is expected, he should follow the linear path, pursue an M.Phil/Ph.D., go abroad, and eventually become an academic with enriched social/

cultural capital. Raj Kumar defied this logic. 'I am tired of the university: its sole emphasis on book learning and intellectualism. Right now it does not appeal to me. I wish to take a break, go back to Chennai—my home town, and engage myself with the Ramakrishna Mission, and participate in its social/cultural activities.' His words carried authenticity. He added: 'Maybe this change would give me a refreshing departure, and make me rethink what I wish to do in life.' I listened to him, and gave him my best wishes. Although I would lose Raj Kumar as a student, I felt happy. I saw courage and conviction; I saw a glimpse of truth.

Why is it that a university like ours fails to attract a sensitive young man like Raj Kumar? An important reason, I believe, is that the university is fast becoming a terribly rational place. Its primary emphasis is on the cultivation of the intellect: how one cognizes the world, classifies and analyses information collected through empirical/ rational categories, and derives logical conclusions. Reason as such is not a bad thing. However, if reason is dissociated from all other equally important faculties like intuition, empathy and compassion, it becomes dry, one-sided and incomplete—

absolutely devoid of inspiration, creative imagination and visionary perception. 'There is too much reading, and that too only complex theories. Believe me Sir, I have almost forgotten to write poetry which I used to do before joining the university.' Raj Kumar's sharp reflections made it abundantly clear: if reason alone becomes triumphant, it is suffocating. Possibly his decision to leave the university expressed his urge to have a breath of fresh air, to find his poetry once again, and to rediscover the world—not with bookish knowledge, but with his own eyes, his own practices and experiences.

Perhaps there is a fundamental problem in the way the university sanctifies a 'knowledgeable' man. He is burdened with knowledge, but not gifted with innocence. He explains, but does not feel. He dissects, but does not relate. As an intellectual he exists only as a critic and commentator. He is not expected to engage himself with simple practices—say, playing with a child and experiencing the same enchantment as she watches a butterfly, talking to an elderly person and getting thrilled at his wide experiences that his wrinkled face reveals, nursing a terminally ill patient and seeing

poetry in his eyes, or just serving clean drinking water to the thirsty children of the street. All these possibilities do not exist because the walls of the university are made of books and theories that only insulate one from the eternal flow of life; its 'field' is only a site for detached observation; its secular reasoning despiritualises life; and its heaviness deprives one of laughter and music. Raj Kumar, it seems, did realise this bitter truth quite early.

One-dimensional rationality, intellect without emotion, burden of books—the university has indeed become oppressive. This oppression gets further intensified as competition—ruthless competition—begins to characterise this learning machine. The result is stress. Professors are competing; students are competing. How to score more, how to grab a research project, how to get a fellowship, how to go abroad: this constant running for a mythical success weakens one's sensitivity. Far from eliminating egotism, careerism and opportunism, the university, it appears, breeds these vices. However, there are soft/sensitive/reflexive minds; they wish to keep their sanity alive.

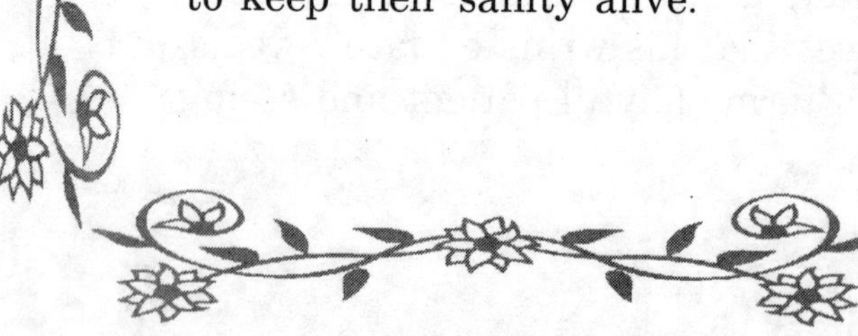

Raj Kumar aroused this hope; and he made me recall yet another incident I witnessed recently. A student, I know quite closely, did reasonably well in her college examination. But then, I saw her crying. She was visibly upset. With absolute patience I sat near her; I was willing to listen. 'My friend—a truly good friend of mine—failed in the examination. How can I be happy when my friend is upset? Why is it that the system is so cruel? Why does it divide people, separate friends?' It was difficult for her to celebrate her success at the cost of someone else's failure. Her anguish was genuine; her tears were sacred.

Raj Kumar's courage, and the young girl's sensitivity, I tell myself, have opened my eyes. Yes, the university gives me my livelihood; I teach, and ask my students to read books; at times, I give them quite a heavy reading list; I take exams, and hierarchise my students. I am no exception; I am part of the same learning machine. Yet, the inspiration that I derive from these rare moments is that I can try, I can innovate; and with humility, prayer and faith I can take a step forward to lay the foundations of a new culture of learning that values books as well as lived experiences, intellect as well

as compassion, analytical reason as well as creative inspiration. It is like having a culture of learning that has not lost its soul. Raj Kumar and that young girl, I love to imagine, would find no reason to leave or feel upset; instead, they would bloom like flowers in my ideal university.

13

My Kurukshetra

Retaining one's sanity, or keeping the flame alive, it seems, is not very easy. At times, all ideals get unsettled; all values become obsolete; all noble aspirations seem like hallucinations. And then, as pragmatic people keep reminding us, the 'real' world—its hard stuff—is overwhelmingly powerful; we are bound to get carried away, lose sanity, and submit before the dominant commonsense. For example, these days how often we are told that corruption is all around. Corruption is a spectacle; corruption is a distinctive feature of the state—its bureaucracy, its political institutions, and its gigantic economic deals. Corruption is in the glitz of the corporate world. In fact, our material culture, our *rajasic* vibrancy, our sports carnivals: nothing escapes corruption—the mysterious flow of money that turns everything into its opposite. Moreover, corruption is everyday: a normal/ routinised practice that characterises all of us—a traffic constable, a police inspector, an income tax officer, a PWD engineer, a doctor, a school teacher. Television, newspapers, urban folklore, gossip in a village tea shop: all channels of communication tell us that the air we breathe is filled with the filthy smell of corruption, and that is what the world is all about. Indeed, every moment we read about corruption, we talk about corruption, we crack jokes on corruption. And then suddenly I look at myself, and realise that I am changing; the negativity has surrounded my existence; the sweetness of temperament has disappeared, the fundamental faith in the goodness of things has ceased to exist. I am becoming bitter, cynical and faithless.

This faithlessness gets further intensified as I experience the aggression of the narcissistic self in a world characterised by the intensity of desire: insatiable desire for techno-material comforts, for power and wealth, for all sorts of symbolic goods ranging from fashion and stardom to erotic attraction.

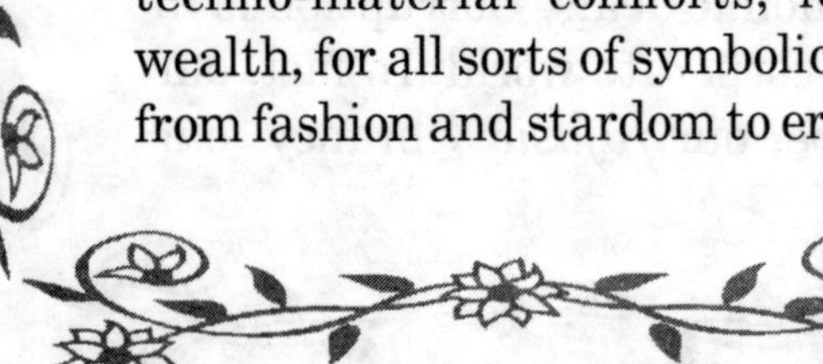

There is a visible demonstration of this narcissistic self. I see it in the streets; I see it in the marketplace; I see it in the university. Capitalists and techno-socialists define it as a symbol of development; neo-classical economists find the possibilities of enhanced growth rate; and sociologists discover yet another index of modernity, and write about new India, urbanity, and the aspiring middle class. And amidst these narratives of legitimisation of the narcissistic self, I get lost. Where is then my longing for a caring/nurturing/relational self: a self that renews itself in harmony, reciprocity and togetherness? I become terribly weak. A narcissistic culture—through its market sites, beauty industries and television spectacles—bombards me. I become voiceless. I fall down.

Moreover, there is always some sort of fear: fear of betrayal, fear of being cheated. How often we tell ourselves that the qualities like honesty, faith and trust do not count in our times. Instead, if you possess these qualities, you are being seen as weak—someone to be taken for granted, a naïve being who can be easily exploited. You come to the street, hire a taxi, remain absolutely polite, don't bargain or mistrust the driver;

but then, he cheats you. You come to the shop, trust the shopkeeper; however, the products you get are of absolutely low quality. The story keeps repeating itself. You expect something, and get its opposite. This pain—this sense of betrayal, this crisis emanating from the breakdown of expectations—is not easy to bear. It begins to harden our consciousness, and we start evolving a counter-philosophy of life: Don't give yourself so easily. Always bargain. Be strict and calculative. Retain distance. Speak the language of power. Demonstrate your authority. Don't allow others to take you for granted.

At times, I too pass through a similar crisis. Am I too soft? Am I terribly simple and naïve: unfit in a world which, as my pragmatic friends remind me, knows only the language of power, calculation and hard bargaining? Should I then become hard, and learn to live with sceptical eyes? These are moments when I get confused; I find no answer; I feel deep pain.

Yet, at these very moments something happens in my consciousness. I begin to rediscover the deeper significance of the epic battle of Kurukshetra. Kurukshetra is not a

site out there; it is in my inner world. All my negativities—my cynicism when I see the all-pervading corruption, my utter helplessness before the seductive presence of the narcissistic self, my fear of betrayal and hence my faithlessness—represent the satanic forces. I, therefore, need to come forward with all my noble aspirations, my positivity, my rhythm to eliminate these forces. But then, who would guide me? Who is my Krishna? Where is my Krishna? Well, Krishna, I tell myself, ought to be within me; it is through my eyes that Krishna would see; it is through my ears that Krishna would listen; it is through my heart that Krishna would feel.

As the battle goes on, a voice enters my ears: Don't complain; instead, create. It softens my soul; it gives me clarity. The practice of creation, I begin to comprehend, does not emanate from egotism; it begins with an awakening of the possibilities each of us is endowed with. This awakening leads us to radiate positive vibrations and do things which are simple, noble and beautiful. The answer to corruption is not to get obsessed with it, but to engage in an act that springs from the spirit of love and cooperation. The answer to narcissism is to lead a life that

seeks to fulfil itself in giving and caring. And the answer to the fear of betrayal is to retain the courage to serve even if obstacles are enormous.

Do these small ventures really matter? Even if I remain simple, honest and faithful, can I alter the world? At this juncture comes yet another voice from my innate Krishna: Change yourself. Don't be preoccupied with the idea of changing the world. Do what you ought to do. Follow your own *swadharma*. And you will realise that slowly, but steadily the world is reciprocating. Once again I find clarity. I overcome my doubt and confusion. Beyond corruption and narcissism, beyond fear and cynicism lies a world that gives me a call. And I find my companions: the silent tree, the flowing river, the radiant sun, the feminine moon, the erect mountain, the infinite ocean, and above all, simple souls, far from being wounded, working in silence, and living with grace and dignity.

14

Even Domestic Work is an Act of Prayer

Here is an assignment my daughter has got from her college. She is required to speak on a paper—written by a feminist scholar—on domestic work and enslavement of women. Out of curiosity I too begin to read the paper; and, yes, the arguments are in order—no deviation from what a politically correct feminist scholar is expected to write! Domestic work remains unnoticed; it is unpaid; and the reason is that in hierarchical patriarchal families married women are compelled to do (even if this compulsion appears to be a form of consent) all sorts of repetitive/mechanical labour—the endless cycle of washing, cleaning, cooking—without being recognised and paid.

It is not that I dislike these arguments. Who on earth would deny the reality of inequality: the way women are exploited, and their possibilities annihilated? Yet, there is something—and something fundamental—I miss in this entire disscourse. Possibly these feminist scholars fail to realise that poetry exists even in a prosaic world, that a man-woman relationship cannot always be evaluated in terms of the conflict between the two antagonistic classes, that even in this bad/ugly world there are possibilies for seeing work--yes, domestic work—as an offering: a gift of love. Not surprisingly, the paper my daughter is reading for her assignment makes me feel a sense of discomfort. Yes, there is an assertion of a politically correct/logically argued feminist critique; but then, how can I deny an equally powerful experience—love/warmth characterising an intimate relationship, and a deep realisation that there is a domain beyond profit and loss, beyond the calculative logic of hard economics?

At this very moment I begin to invoke my mother. She worked day and night, nurtured the family, and brought us up. She was our home—our ultimate shelter. Was she wounded, humiliated and defeated?

My guess is that she was not. Well, feminists might argue that she was a piece of 'construction'; she was trained to accept her exploitation as natural. Although this logic is tempting, I find it difficult to accept. Her alertness, her brilliance, her hold over all of us, and her patience and endurance—in fact, everything of her being—indicated the richness of her *swadharma*. To regard her life as a failure—a case of passive submission before the male authority, or to look at her as a victim of 'false consciousness', or to lament that economics does not calculate the monetary value of her work is to degrade her.

I do not wish to measure the unmeasurable. I believe that it is a fundamental mistake to apply the logic of market economics in a domain that has a quality of its own. My mother taught me this lesson. And I learned it once again as my wife came into my life. Buying vegetables, boiling milk, cooking, washing clothes, getting the tiffin box ready for the school-going daughter: there was no end. Yet, the flow of love, trust and care gave a new meaning to all that she did. No wonder, it was in the kitchen that we discussed poetry; the vibrations of the washing machine did

not prevent us from engaging in a discussion on Ivan Illich and Rabindranath Tagore. Even when there were moments of physical pain and fatigue, she did not bother to calculate whether she could have used her time in more 'productive' work, and earned some money. A piece of work performed with the spirit of love becomes divine; one does not feel like trivialising it by calculating its economic value. Her kitchen was our temple, her simplicity was our prosperity, her work was our lesson of detachment, and the fact that , despite the recurrent cycle of domestic work, we could continue to dream was our ultimate reward. I may be accused of overstating my subjectivity. However, for me, objectivity ceases to have a meaning if it is not subjectively experienced. In the world of 'facts' my experience too has its legitimate space.

My mother, my wife—both left the world. And hence domestic work, for me, is no longer a theoretical domain; I am destined to get involved in it. Yes, the feminists I am referring to might ask me: 'Till yesterday you were romanticising women's work in the family because you did not participate in it. But now as you are getting your hands dirty, is it still possible for you to

romanticise?' My only answer is that domestic work is qualitatively different; it is beyond the logic of payment; and moreover,every piece of work—yes, even cooking/cleaning/washing, when performed with the right attitude, takes us to a higher domain of existence.

See how my day begins. I get up quite early. I know I have to be active. Storing drinking water, organising the kitchen, buying vegetables, planning the lunch menu, cooking...the work does not end. Calling the electrician for repairing the faulty plug point, making a telephone call to the gas agency for booking the LPG cylinder, paying the telephone bill—I see myself engaged in a series of activities. True, there are moments when I get tired; I lose my temper. But then, even my university work makes me tired. That is normal human weakness. The real truth is that domestic work is not dull, monotonous, meaningless. Why should something that is so fundamental for our everyday living be considered trivial? Why should everything need to have a visible public character? Why can't some work be done in silence? Why should we always complain that domestic work is not recognised? In fact, the more we look at it

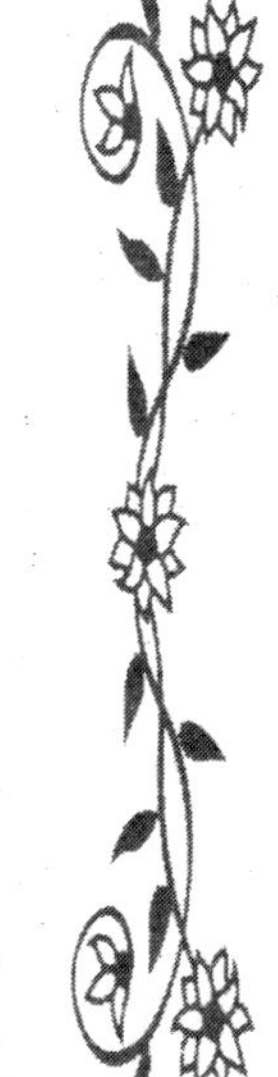

with negativity the more burdened we feel. Instead, everything that is necessary for retaining the warmth of a home ought to be done with joy and creativity. It does not matter whether it is being 'recognised'. It is beautiful in its own right. It is enabling, not constraining. Engaging in a philosophic discourse or delivering a meaningful lecture or writing a book, I realise, is not necessarily in conflict with my domestic work. As a matter of fact, I become more integral, more rooted, more earthly. The entire attitude to work begins to alter. Work becomes an act of prayer, an offering, an expression of love. And possibly for the first time in my life I realise the messages my mother and my wife used to radiate. In their absence they become truly alive. They become my sources of inspiration. I find my poetry. And I acquire the courage to see beyond economics—beyond soulless feminism.

15

For My Own Quest

Living amongst experts seems to be a necessity in our times. Experts, it is said, are bound to be everywhere. Because knowledge is complex, skills are specialised, and the division of labour is minute. And hence you need a special mechanic to repair your television set, a doctor to heal your wounded body, a psychiatrist to make sense of your sleeping disorder, a career councillor to resolve your child's confusion, an interior designer to decorate your room, a travel coordinator to plan your holidays, an event manager to arrange a party at your home, and a dietician to decide your food. Experts invade almost every aspect of one's life, and this invasion is considered to be legitimate, progressive, a token of modern living!

Living, I admit, is interdependence. It is good to be humble, and realise that one can't know and do everything; there is no harm in seeking the help of others. This delicate art of interdependence nurtures the culture of reciprocity. However, there is a difference between the culture of reciprocity and what is happening in our times in the name of our reliance on expertise. We forget all limits; we lose our own potential and confidence; we get completely paralysed. There is no sweetness in the relationship any more. Now it is all about passive consumption. For everything, be it food or culture, travel or pilgrimage, aesthetics or spirituality, we pay, hire the services of experts, and consume their prescriptions.

However, there are moments—and I adore these moments—when I feel like revolting. I wish to be alone. No dependence on experts. No mediation. Between me and the universe—I would not allow anybody to guide, prescribe and restrict the flow of my prayer, my music, my intimate conversation. Let the experts dominate our hospitals, industries and universities. But in my most pure and sacred domain of aesthetics and religiosity I do not need them. I do not need a professor of literature to tell me how to

read William Blake; I do not require a specialist in divinity studies to interpret the *Bhagavad Gita* for me; I loathe the idea of hiring a travel expert to see and feel the Himalayas; and believe me, I do not need yet another academic text on Gandhi and Tagore to tell me what to learn and unlearn from them. I need peace; I need silence; I need the purity of my experience.

Let me begin with the narratives of some of those extraordinary moments. The other day I saw and felt once again the continual play of life and death. A retired professor had died; there was a condolence meeting in the university. However, at the same time there was yet another meeting taking placc: this time a professor—full of energy, fame and at the peak of his career—delivering a lecture amidst the packed audience. The simultaneous enacting of these two events took me to an altogether different domain. Everything began to seem like a play—death and life, collective mourning and vibrancy in a lecture hall. The professor who had died was great; but then, nothing should stop, the game must go on: lecture, philosophic argumentation, tea, snacks... It was at that moment that I heard a voice emanating from the infinite: 'See it as an ongoing play, accept

it as a cycle of *samsara*. Death is normal, life is normal, a condolence meeting is normal, and an alive discussion in a lecture hall is normal. One stage leads to another. Everything is in flux. And hence don't get unduly attached to it. See the game. Participate in it. But do all this with an awakening. Your ego does not matter. Yesterday he was lecturing, today people remember him at a condolence meeting. Now someone else is lecturing; tomorrow, who knows, there might be another condolence meeting. With this awareness comes detachment. Detachment is not a retreat into passivity; detachment is active engagement with clarity—living, participating, yet existing as a spectator observing the game in silence'. Yes, with this voice I did find my *Bhagavad Gita*; I rediscovered its lessons of detachment and karmayoga. No expert. No philosophy professor. No scholastic commentary. It was my own, intimate and hence very real *Bhagavad Gita*.

I grew up in Bengal where the iconisation of Rabindranath Tagore has reached an absurd level. At every locality you come across experts on Tagore—his music, painting and literary creations, his

relationships, letters and even foreign trips. Almost everything relating to his life and death is subject to expert gaze. However, this heavy load of information causes a sense of bondage; I begin to feel restless; I wish to come out of these intellectual deliberations and contestations. I feel like seeing, touching and experiencing the poet in my own way—absolutely natural, simple and rhythmic. And I begin to find him in the abundance of nature; I listen to his prayer as I look at the vast sky full of stars and planets. With a sense of wonder and gratitude I sense the meaningfulness of my existence in this beautiful universe. I feel his restlessness when I encounter the wall of barriers: be it the regimentation of schools as learning machines or inflated egos of mighty nation-states. Only then do I realise his notion of 'surplus' as man's religiosity: the quest for seeing the infinite in the finite, and experiencing the finite as an articulation of the infinite. I talk to him in silence. There is no burden of scholarship—no burden of 'right' information, 'right' interpretation! My Tagore is not the Tagore of experts, my Tagore is like the upward flame of *agni* (fire) taking my offering and prayer from the temporal to the transcendental.

There are moments when I also find my Gandhi. Nathuram Godse could not eradicate him. And believe me, I have seen him; I have talked to him. But before that let me state the obvious. We live amidst the culture of indulgence—the culture that stimulates our identities as consumers of symbols, commodities and technological spectacles. We become dependent on the outer show. This leads to the intensification of desire for having, possessing and conquering the world. And then a question confronts me: Is it freedom? Or is it my dependence on the externalities of life? Is it that I am nothing without my car and laptop, my mobile and credit card, my air conditioner and television set? At this very moment Gandhi appears, and whispers into my ears: 'Evolve your inner resources, your soul force. Only then can you be free from insatiable desire and indulgence. You will become calm. You will look at people and nature in a more harmonious way. That would be your real strength and courage—the beginning of ahimsa.' Gandhi becomes very real and alive. And hence at this deeply intimate moment I need not bother how Ambedkarites, Marxists, post-colonialists and innumerable other experts look at my interpretation.

Let there be some domains where I remain free from the trap experts create through their knowledge. Let my experiences—simple, genuine and spontaneous—continue to enchant me. Only then, as I tell myself, is it possible to be free from the burden of 'science', and echo William Blake:

> To see a world in a grain of sand
> And a heaven in the wild flower
> Hold infinity in the palm of your hand
> And eternity in an hour.

16

Horizons Beyond the Research Committee

The other day as a member of a research committee I was listening to a Ph.D. scholar. She was explaining her synopsis on the phenomenon called new religious movement: the overflow of gurus, their disciples, and the discourses on yoga, spirituality and healing. A theme of this kind, needlessly to add, aroused immense interest, and generated a lively debate. An eminent expert—with a highly developed social science imagination—did not forget to remind the scholar: 'These gurus are highly charismatic. And hence be alert and careful. Don't get carried away. Don't lose your criticality.' A series of sociological insights were put forward. Let me recall some of these observations.

First, it is the all-pervading logic of commodity fetishism. Religiosity seems to have lost its deeper meaning; it is presented as an attractive package. In the supermarket of religiosity there are multiple choices, and depending on your need and financial capability, you can buy any of these packages. Second, we live in an age heavily dominated by television images and visuals. Television creates needs, generates spectacles, produces stardom, and induces us to believe that all that appears is bound to be true. As a matter of fact, many of these gurus, as the proliferation of 24*7 religious channels suggests, have been 'constructed.' Their continual appearance on the screen makes us believe that they are real and authentic. And third, India, it seems, has changed; new values, new aspirations and new anxieties have occupied the cultural landscape of the middle class in the era of neo-liberal global capitalism. They are mobile and prosperous; yet, they are insecure and full of fear—fear of failure, fear of lagging behind, fear of illness, and fear of stress. Essentially, as it is argued, these gurus target the middle class. They emerge as new age psychotherapists and healers; they are sufficiently smart to prescribe what makes sense to this class: retreats and workshops, art of living classes, yoga camps, and ayurveda with a new package. In other words, as this critical analysis suggests, there is no such thing as religious ecstasy; everything is need- based; religious gurus are sellers; we are buyers. As the logic of sociological explanation triumphs spirituality loses its magic.

All these arguments are superb. I see their relevance. Yet, a question continues to haunt me. Is it then the ultimate fact that we are destined to live in a world in which there is no depth of spiritual longing, a world in which everything is sociologically explicable? Is there nothing beyond these religious gurus, beyond 24*7 television channels, beyond the middle class obsession with instant solutions to all problems, including the psychic/existential ones? The fact that I am asking this question indicates that my longing is taking me beyond the horizons of the research committee. Possibly a journey has already begun...

In the process of undertaking this journey one thing is becoming clear to me: truth transcends the seeker of truth. Even an extraordinarily gifted seeker has an

embodied existence, and as a result, the ego, despite all sorts of *sadhana*, may not wither away; it is, therefore, not impossible to find some sort of asymmetry between the radiant truth and its seeker. However, this human failure does by no means belittle its beauty, purity and grace. Even if all our poets, saints and mystics have sinned, the fact is that the sun is still rising and giving its warmth, the flower is blooming and radiating its beauty in symmetry, the mountain is erect and indicating its expansion and silence, and the sea is continually manifesting its infinity. So why should I let myself down, and think that there is no truth? Truth is limitless; it is nobody's monopoly.

There is yet another revelation. There is no God in temples, no mystery in discursive discourses, no realisation in the mechanical memorisation of scriptures. Instead, seeing is experiencing; experiencing is believing; believing is internalising; and internalising is awakening. And when it happens, truth does not require any proof, any lab experiment, any logical demonstration. The sea is vast, the tides are huge, they are approaching the seashore, and they are crumbling—the way our inflated egos crumble; what remains is the eternity of the

sea—its infinity. When we see it not just through our eyes, but through the entire being, we realise that our liberation lies in merging with the infinite. The guru is inside; everything is inside; and when seeing becomes truly enchanting, the inner and the outer merge. That occasion is noble, pure and sacred.

I am not sure whether my sociology, my role as a member of a research committee can take me to that moment. Possibly it cannot. However, I am more than what my sociology can describe. I am a humble seeker undertaking a journey. And in this journey there are moments when I get the glimpses of what I am striving for.

I am at Dhanaulti—a small Himalayan hamlet. It is 9.30 pm. The sky is vast, clear, extraordinarily beautiful. There are pine and oak trees trying to touch the sky. Amidst that intense silence something happens. A tree begins to whisper. My body trembles; I get myself dissolved; I become nothing. In that nothingness I become everything. I become the tree singing and praying since eternity. I become the sky, the mountain. I become eternal silence. My ego, my sense of separation/exclusion, my fragmentation—

everything becomes shallow and meaningless. A sense of connectedness begins to fill me with joy, with an experience of abundance. At that moment it does not bother me whether our religious gurus are doing business; it does not matter whether the urban middle class is buying religious packages. What matters to me is the silence of the night, the whisper of the tree, its connectedness with the distant star, the mountain as a witness, and my ego becoming futile and meaningless. I get my yoga—my principle of connectedness. Who says that it does not exist?

It is a Durga Puja celebration. The site is a famous temple in South Delhi. In order to attract and involve the devotees the organisers have arranged what is known as the Musical Chairs Contest. As the music begins, the participants are required to move around a circular path; and then suddenly the music stops, they are now expected to sit on the chairs which have already been placed around the path. However, there would be a scarcity of one chair, so one participant would not be able to get her seat, and she would be out of the game. The process goes on, and in each round one participant would be out. I keep observing the game with great curiosity. And suddenly a flash of truth strikes my mind. Life is like this sport: a constant movement in this cycle of *samsara*, and always someone disappears or withdraws from the game. Nobody knows when one's turn will come; it is not predictable; and that is why it is a sport. However, one thing is certain. Everybody has to go. This makes me see the meaning of death. Death is like that disappearance from the game. Death is certain. Yet, nobody knows when it will come. So what do we do? Should we be perpetually afraid of death, and hence fail to participate actively in the sport of life? Or should we be in a state of denial, and think that it would never happen to us; it happens only to other people? Beyond these two possibilities lies yet another domain: truly fulfilling and meaningful. We continue to participate in the sport of life without being saddened by the fact that death can come any moment; we realise that this obsession or fear deprives us of experiencing this very moment, its aliveness and its totality. However, at the same time we remain awake while playing the game; so when death comes we do not get surprised and unduly shocked; instead, we accept it as inevitable—a rule of the game, the way the

enthusiastic participant of the Musical Chairs Contest would leave the show with a smile when her turn comes.

Let the journey continue. Even if I fail time and again, let me arise. Let me pray: 'I have got your glimpses. Reveal yourself again and again. Reveal yourself at every occasion in my life: when I teach, when I cook, when I read, when I engage in the phenomenology of everyday life, when my body trembles and becomes weak.'

And then I saw her in my dream. She stood on the top of a mountain. She smiled with the freshness of the full moon. I forgot my academics, my reasoning. I started climbing...

17

On Bio-data and Final Laughter

'I find it difficult to write my bio-data.' Shephali's words surprised me. Shephali is a young/vibrant student engaged with multiple activities. No wonder, we all expect her to think of her career profile: fellowship, job and achievement curve. At this stage of her life, as it is expected, she must project and assert herself, and master the skill of writing her bio-data; only then can the concerned authorities be convinced of what she is. We all take it for granted. We write about ourselves: our degrees and diplomas, our awards and achievements, our publications and foreign trips, and even our politically correct radical/grassroots activism. Shephali, I believe, knows all this—how in this hard/pragmatic world filled with the ethos of reckless competition one must continually demonstrate one's capabilities. Yet, as she says, it is difficult for her to write about herself. In fact, in her discomfort I see a meaning; I am inspired to rediscover the truth of existence.

We live in a world that stimulates one's ego; living itself is centred on the rise and fall of the ego, and resultant pleasure and pain, hope and despair. What is the nature of this ego? It equates my essence with my bounded existence: my body, my wealth, my intelligence, my education—the feeling that as if I were really important, and I matter. The more 'successful' I am the more inflated my ego is; I carry the entire baggage of my rankings, awards and achievements; and my bio-data becomes thicker and thicker. However, this ego is insecure; it is perpetually afraid of losing what it has achieved; and hence it cannot exist without asserting itself. Someone has better publications than mine; my ego gets wounded. Someone does not recognise me at a meeting; my entire day gets disturbed. And someone says, 'You are really great;' I begin to fly. This restlessness/insecurity is further intensified because the world has now made everything into a saleable commodity. Promoting/advertising oneself seems to have become the norm of the day. 'I did my Ph.D. from Oxford. I worked with the adivasis of Chhattisgarh. I organised rural women in an anti-liquor campaign. My book has been published by Routledge'. How often one comes across a bio-data of this kind in the academic world. True, it is seen as normal and desirable; it is taken for granted. Yet, the fact is that it is also an act of selling oneself, albeit in a subtle form. A beauty queen in the glamour world sells her physical beauty; a progressive academic sells his intelligence and political correctness. Shephali, I feel, is immensely sensitive; she sees it clearly; it pains her.

If the act of promoting one's bio-data is like celebrating one's ego, is there anything beyond it? Possibly it is emptiness: no burden/armour of a separated, proud or wounded self, but only an all-pervading void. Breaking the armour or annihilating the ego is like having an experience of the lightness of emptiness. There is no baggage, no burden of past achievement and failure to restrain me. My bio-data is my past; it is dead. My emptiness is my 'now and here'; it is alive. And hence I receive; I sing; I dance; I enter into ecstasy. I am no longer a professor with

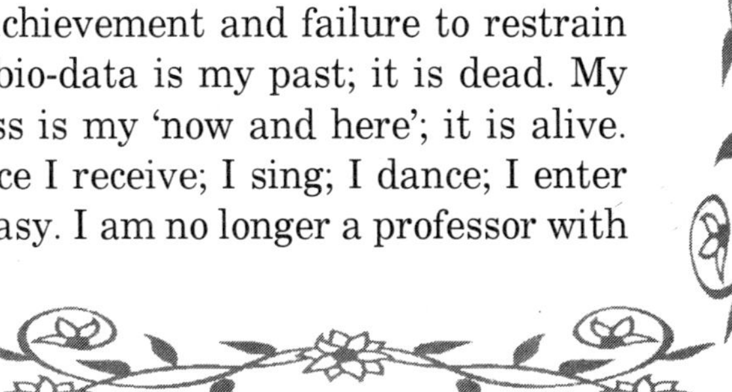

a heavy burden of knowledge and publications; I can get dissolved into the welcoming smile of a rural peasant. I am no longer a powerful man protected by the mighty state's armed forces; as the train stops at a lonely railway station, I can get down, eat *samosas* at a local tea shop, and play football with school children. I become free; I become receptive and communicative. Indeed, with this emptiness comes the realisation that I am beyond all the parameters of a technically perfect bio-data. I am essentially a manifestation of the energy that moves around this all-pervading emptiness. My emptiness is not nothingness; instead, it is my real substance. With the expansion into the infinity of emptiness emerges love and compassion—freedom from fear, insecurity, pride, ego, failure and success. It is *ananda*.

Shephali, it seems, is striving for *ananda*; and the very idea of writing a bio-data diminishes its beauty. But then, it is also true that in this phenomenal world one's bio-data is seen as an indicator of one's orientations, inclinations and aptitudes, and it is in this sense that educational institutes, research organisations and employers would require it. Possibly Shephali—so long as she needs a job or a fellowship—cannot escape this logic. Yet, there is a possibility. She can see and feel the entire exercise as a play; she need not get excessively attached to it. Instead, she can elevate herself to a witness looking at this entire play, laughing at it, and not getting oppressed by it. It is the wisdom to laugh at oneself, to laugh at one's awards, achievements, degrees, rankings and positions. With this ultimate laughter things become empty once again.

It was at Mukteshwar—a small Himalayan hamlet—I met a sadhu. He spoke of silence and fasting. He looked at the vast mountains, and laughed. 'I am not different from all that you are seeing around. One day I will get dissolved into this abundance of nature.' His words took me to a different domain. I could not acquire the courage to request him to narrate his bio-data.

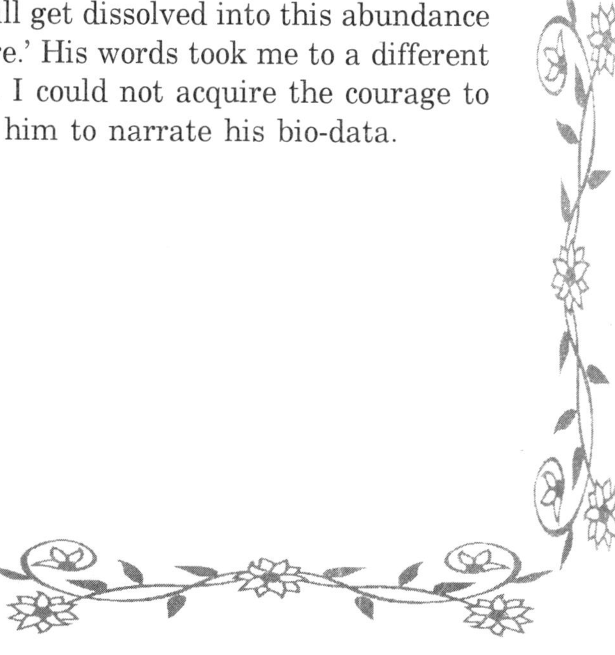

18

From Revolution to Rhythm of Living

Growing up with the idea of revolution, for our generation, was quite normal. In fact, as I recall, even during our childhood we heard of revolutions—from the French, American, Russian, Chinese revolutions to our own revolutions in the making. Marxists, Leninists and Maoists, we were told, would arouse the peasantry and the working class, and transform our society. From feudalism to capitalism to socialism—the idea of revolution was everywhere: in books, in political slogans, in progressive theatre and in radical utopias.

But then, despite this heavy dose of revolution, I have not yet understood what this total change is all about. Every morning the sun rises, people go to their workplaces, and parents pressurise their children for studies. Every day babies are born and doctors issue death certificates in hospitals; youngsters fall in love and lawyers file divorce petitions in courts. Every day there is laughter and joy, pain and separation. And in the silence of every night there are new promises as well as broken dreams. Where does revolution exist amidst this recurrence of everydayness?

The proponents of revolution, I know, would tell a different story. Revolution is a big event—magnificent and spectacular. It leads to total change. The entire politico-economic structure gets altered, and what emerges is a new culture. Time is linear; revolution would take place in future; and on that eventful day we will all change. We would no longer beat up our wives, take bribes, accept dowry. We would not exploit others. We would become egalitarian, peaceful, communists par excellence! We would not be the prisoners of everydayness. However, that grand day does not come. Generations after generations, as every morning we wake up, we confront the same world: the recurrence of everydayness as well as what newspapers publish: violence, rape,

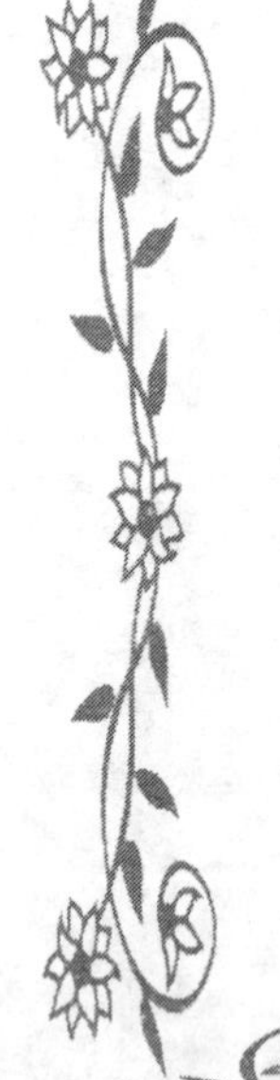

murder and corruption. It is, therefore, not surprising that there are moments when the proponents of revolution get disillusioned . You expect something big, and find very little in return. From euphoria to despair—yesterday's revolutionary becomes today's cynic!

What then do I do? Do I live with the idea of a hypothetical tomorrow—the final day of revolution? Or do I live with despair and cynicism? None of these options are mine. Possibly at this stage of my being I have begun to see life—its everydayness, its 'here and now'—as a rhythm filled with possibilities: not big and spectacular, but immensely meaningful.

This rhythm is sensitive to the activities that take place every day in a network of spatial locations ranging from bedroom to kitchen, workplaces to shopping centres, and streets to neighbourhood. These small domains are not unimportant. Even if 'determined' by huge politico-economic structures, the fact is that it is only in these domains that we live every moment, and realise what we are, what we wish to be and what our possibilities are. All our big designs (fight global capitalism, bring about socialist revolution; fight the façade of parliamentary democracy, introduce people's communes) would remain merely an abstraction, if we remain indifferent to the space we are familiar with, the space we know the best. I do not know whether one day capitalism would be abolished; but it definitely makes sense to me if I refuse to be persuaded by its culture industry that promotes the ethos of conspicuous consumption. Even if I cannot make any difference in the brutal practice of violation of human rights, it is still important for me to respect my maid as a human subject, talk to her the way I do with my colleagues, and give her due wages. How I live here and now as a father/husband/neighbour/worker (or how small things matter even if big events remain elusive) defines my rhythm. True, one can say that I am nothing; the 'system' is everything. However, at a deeper level the 'system' is nothing apart from human practices. You are the system. I am the system. Capitalism is not something that exists only out there. It is also our greed, our insatiable desire for consumption, our insensitivity to gross inequality. Patriarchy is our burden of masculinity, its insecurity and fear of the feminine. And hence there is no magical

revolutionary or a political machine annihilating the foundations of patriarchal capitalism from our consciousness. Each of us has to alter the system that exists within. That is the rhythm of life. At Chail—a small Himalayan hamlet in Himachal Pradesh—there is a beautiful temple on the top of a mountain. It is cold, we are shivering, and then a worker appears with a kettle full of hot tea. His offering gives us warmth and energy. Out of my urban mannerism words come out of my mouth: 'Thank you very much for your tea'. 'Who am I to offer you tea, Sir? It is He who does everything'. His simple/authentic words get dissolved into the vastness of the mountains and clouds all around , and suddenly the self-promoting culture of narcissism vanishes. He is neither Lenin nor Mao. He is merely an unknown existence like a tiny flower radiating and reminding us of the rhythm of life.

This rhythm, it is obvious, does not mean that you keep waiting for a miracle to happen one fine morning. Instead, it is a matter of everyday practice, a continual flow of life. It is like seeing possibilities in the ordinary. You are not a Gandhi. You are not a Marx. Nor are you waiting for them to emancipate you. You are a clerk in a bank, a teacher at

a primary school, a cab driver in a metropolis. However, the domain of your work can act like a zone of possibilities. Imagine you work with joy, and make an extra effort to help a new customer to open her account in your bank. Imagine you come to the classroom with fresh ideas, love children, act like a catalyst, and arouse their interest. Or imagine even when it is late at night your cab is ready, and you do not hesitate to take a newcomer in the city to a remote colony. Only then can the mundane become sacred. With this rhythm of life abstract theorisation (Do I have 'agency'? Am I working only in the domain of the superstructure? Am I undermining the larger structural issues?) becomes secondary. What matters is the willingness to begin—the ecstasy of doing. I have not seen revolution. But I have come across simple people with extraordinary stories, and their rhythm has touched my entire being.

This rhythm is silent; it is a process of inner transformation. No machinery can get it done for us. In fact, even when we constitute a political party or a trade union, it becomes truly vibrant and meaningful only when we see ourselves as enchanted souls, when the inner calling gives a new meaning to the outward activity. Look at, for instance, our insatiable greed. You can keep debating whether it is greed that gives birth to capitalism, or whether it is capitalism that induces greed. One who leads a rhythmic life does not debate so much; he acknowledges greed as a source of violence, exploitation, envy and perpetual restlessness; with this awakening he goes deeper, and finds his ultimate treasure: the contented self that finds fulfilment in reciprocity and symmetry, in giving and sharing, in love and compassion. And this journey becomes the most effective resistance against the lure of capitalism: more powerful than all the slogans, all the demonstrations put together. Likewise, look at fear—the fear emanating from insecurity. It is the fear of losing what the ego regards as its possessions; it is the fear of the other. The male patriarch is afraid of the feminine; he seeks to subdue it. The fascist is so insecure and fearful that he represses the other. And behind every war lies the fear of the enemy. To lead a rhythmic life is to understand this fear, and then overcome it through a deep realisation that one is not insulated from the universal energy that flows through everybody. Small, but meaningful actions—meeting people,

sharing joys and sorrows with them, and unlearning all sorts of artificial decorum—help one to overcome insulation and associated fear, and realise that the mind is indeed an ocean. Love does not emerge at the end of revolution; love is the beginning of everything.

I do not know whether Marx's communism or Gandhi's Ram Rajya would ever become a reality. But then, I am sure, there are people amongst us who do not wait for a revolution; they lead a rhythmic life; they see poetry even in a prosaic world, find music amidst noise. Yes, I admit, for many of us, this rhythm is not easy to continue. One disturbing question that is bound to confront us: would the world change if I change? Possibly it would not. And herein lies the significance of prayer: prayer as a mode of healing, a cleansing process. Revolutionaries may be scientific and secular—completely dissociated from the transcendental, and hence they are also the ones who can be easily frustrated. But the rhythm of life cannot sustain itself without deep prayer: the way every piece of good music is a longing for the divine. Even if your honesty is insulted in a world of scams and scandals, and your love is not

reciprocated in a culture of bomb blasts and police firing, you should not feel defeated. Let us acquire the strength to realise that our rhythm is our truth, and God appears when a fountain is created in a desert. Let life fulfil itself in the purity of simple/beautiful endeavours.

19

Silence is Purifying

I am engaged in a profession that demands the ceaseless flow of words. I come to the classroom, deliver lectures, and spread out words. Words breed words. I am in a seminar, I begin to speak, and I am conscious that I am required to interrogate others, assert my points, provide arguments and counter-arguments, and charm the audience through carefully crafted words. I am a social scientist, and hence I begin to believe that it is my historic responsibility to speak of Bourdieu and Foucault, Lyotard and Habermas; I must acquire the capacity to speak endlessly on cultural construction, identity politics, gender discourses, media simulation, globalisation, neo-liberalism and multiculturalism. Not solely that. My words—particularly if I wish to make my presence felt in the appropriate academic circuit—must always express the language of political correctness: Gandhi is wrong, Ambedkar is correct; nationalists have deceived, subalternists have revealed; unity is an imaginary construct, conflicting differences are truly real! In my profession, it seems, there is no escape from words: carefully planned words, mystified words, words as capital. I become a prisoner of words.

Yet, there are moments—and these are indeed revealing moments—when I feel the superficiality of the game of words. Words have become an obstacle; words are taking me away from the realm of light and truth; I get tired of words; I feel like forgetting, unlearning, throwing all my lecture notes/ seminar papers into the dustbin, and becoming pure and innocent once again. And at these moments I realise the worth of what a sadhu at Mukteshwar once told me: 'Believe it, silence is your truth, your virtue'. Silence acquires a new meaning. Silence is not escapism. Silence is not withdrawal. Silence is calmness that helps one to realise the

depths of existence. It is in the domain of silence that life and death merge, and energy becomes a rhythmic flow of love rather than aggressive principles of domination. It is silence that enables us to listen to the finest music, and write our epic poetry. Silence purifies us. Words cannot describe what we experience in silence. Words are calculative, strategic, diplomatic, ornamental, politically loaded; but silence has no 'ism'; it is unconditional; it is like an all-pervading void—eternal and immortal.

As I enter the zone of silence, I become free. There is lightness. My 'ego' disappears. There is no need for providing arguments and counter-arguments. Nor am I required to explain, theorise, legitimate and deconstruct. I need not bother whether anybody is understanding me. I am free from all these anxieties. Instead, I am swimming and flying. I am at Kaudiala—an unknown place at the foothills of Shivalik. The river is flowing; the mountains are witnessing this wonderful rhythm; the sun has just begun to disappear; it is time for dusk; and like a small particle I am also an integral part of that sacred moment. There is silence. It invites. I can now listen to the story the

river tells me: the mystery of life, the meaning of a journey. The mountains tell me what it means to be a witness—erect and stable. And then I look at the sky, and I find Jibananda Das: a great poet I have always adored. He comes down from the distant stars, and whispers into my ears: ' In that amazing night the stars who had died in the sky's lap thousands of years ago came…bringing with them countless dead skies. The beauties I had seen dying in Assyria, Egypt, Vidisa seemed to be ranked on the distant frontiers of the skies. Was it to trample over death? Was it to express the deep-felt triumph of life? Was it to raise a stern fearsome tower of love'?

What does it matter whether others are understanding me or accusing me of my metaphysical madness? The fact is that at that profound moment of silence I am contented. I realise my truth. I break the wall separating the phenomenal from the transcendental. My vision becomes my reality. No wonder, it is only in silence that she comes, heals my wounded self, and assures me: 'Believe me, I have not gone anywhere. I am your breath. I am your smell. I am your inspiration—your energy'. Again, in silence I realise that Gandhi walking through the bylanes of Noakhali is like Jesus on the mount delivering his sermons. Jalaluddin Rumi becomes alive once again. I feel his touch. He inspires me to sing:

> Through love, illness becomes health.
> Through love, a curse becomes a blessing.
> Through love, the thorn becomes a needle.
> Through love, the home is lit up.
> Through love, the dead man becomes alive.
> Through love, the king becomes a slave.

Silence is indeed purifying. It softens me. It frees me from the burden of discursive discourses. It rescues me from the aggression of words: their continual bombardment on human consciousness. Yes, I know, I live in the world amongst shopkeepers, bankers, brokers, mathematicians and professors; and hence there is no escape from the continual flow of words—words of trade and commerce, science and reasoning. But then, since I have tasted the music of silence I acquire the confidence to believe that it is possible to face the world, and even amidst its continual noise one can remain calm. Or possibly words—the words I have to use in my everyday as well as professional engagement—begin to acquire a new

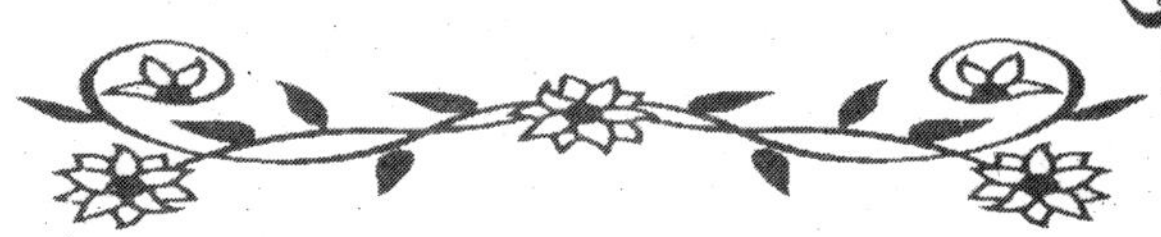

meaning, and what the Buddha once regarded as the virtues of 'right speech' make sense to me: 'Undisturbed shall our mind remain, no evil words shall escape our lips; friendly and full of sympathy shall we remain, with heart full of love, and free from any hidden malice'.

20

On Pain and Compassion

There are moments when I get carried away by a strange feeling: one is inherently lonely; even if there are friends and relatives to celebrate festivities and carnivals of joy, there is no one to accompany one when one suffers, feels acute pain, and experiences psychic trauma. With suffering begins the crisis of existence. One confronts a series of questions relating to the deeper meaning of life, faith, friendship and relationships. Imagine the crisis resulting from the loss of the dearest ones, or acute bodily pain/physical disability, or inexplicable tragedies like accidents. This suffering appears distinctively unique, and one fears that no one can understand or share it. It is you who have lost your mother/wife/child.

No one can feel it the way you do. Because, for them, it is just another death: a normal/ routinised event taking place all the time, something history would never bother to remember. Or, maybe they would spend some time with you, and eventually go back to their own world after consoling you: 'Time is a great healer. Have patience. Take care. Everything will be in order'. But you alone know that things would not be the same again. You alone know what it means to live without your loved ones. For you, history, politics and culture cease to have a meaning without them. Because you alone experience the absence of that touch, that smell, that journey. And you know that you cannot explain it to anybody. You are destined to bear it alone. Or maybe, you come near a tree or a river or look at the distant star, and cry in silence. Or imagine yourself hospitalised, your body is in acute pain, and doctors have predicted that this diseased body is decaying fast, your days are numbered. Yes, your spouse, your parents, your siblings, your children, your friends: everybody is concerned. They are arranging money, buying medicines, consulting doctors, visiting temples. But you alone know what it means to be diagnosed with a fatal disease;

the pain in your body is your pain, no one can take it away from you. At midnight when everybody is asleep, you look through the window, enter an unknown territory, and tears fall from your eyes. You alone know what it means to be objectified when doctors reduce your body into an experimental site, and bombard it with all sorts of medical gadgets. These moments are moments of terrible isolation. It is not easy to retain faith. Instead, there is every possibility of faithlessness afflicting your consciousness.

Possibly you feel a sense of injustice—some sort of victimisation. 'Why me? What wrong have I done? Why am I suffering so much?' It is not easy to cope with these questions. You get angry with God, with everybody around. Friends and relatives—all appear to be privileged ones far remote from the world of your pain and suffering. You see yourself condemned—a victim of bad karma. Or you lose faith in everything and become utterly nihilistic. There is no meaning in anything; your suffering further reinforces the purposelessness of this chaotic/arbitrary world!

Moreover, you tend to lose faith in others. They don't bother to understand you—you

lament. This crisis gets further intensified in modern times. Modernity intensifies our speed, mobility and achievement-orientation. No wonder, in modern times, despite the abundance of technological comforts, we live with a sense of scarcity—scarcity of time. Everybody is running; everybody is in a hurry. Where is the time to stop for a while, and look at the tormented self with compassionate eyes? In a way modernity has disempowered us; it is becoming increasingly difficult to have patience, to remain calm and stable, and listen to others. So even when we have all the externalities of life, at the moment of crisis the world seems deserted. No dew drop, no flower; only dust, endless dust.

Is it possible to come out of this grave moment, and restore faith? I ask myself this question time and again. The mind gets disturbed and agitated; there is fear, and I tend to fall into its trap. Yet, there are moments when a flash of truth illumines my consciousness. Suffering, I begin to realise, is not just mine; suffering is universal. I see it everywhere. Hospitals are over-crowded; newspapers are filled with obituary columns; and anguish/pain is all around. I admit that an awakening of the universality of suffering is no consolation; nor does it diminish the intensity of pain. Nevertheless, it becomes easier to free myself from the obsession with my own suffering; I acquire the strength to see beyond myself. What is most important is that it leads to the acceptance of what is inevitable: all that is temporal is bound to decay, die and decompose; this body is a source of pain, death is pain, separation is pain. Learning to accept the inevitable is a great art; it gives one the perspective to see things clearly, and live life without undue apprehension and anxiety. This is like accepting the inevitable with grace.

Another thing happens. As I realise the universality of suffering I stop blaming others. Instead, I acquire sanity and realise that there is no difference; we are all fellow travellers passing through the peaks and valleys of life-trajectory; and we need one another. With this awakening, I feel, begins a paradigm shift in our consciousness. Instead of becoming angry with others, we listen to them, and in the process we too are heard. We realise the depths of sharing and listening, calmness and tranquillity, and rethink the terrible speed of our age. Pain—in its purely physical sense—does not disappear. However, it is pain that makes

us familiar with our potential Buddha nature. Love begins. Faith is nothing but the awakening of love. Yes, I have seen it. She was in acute pain; a young doctor conducted the bone marrow test. Yet, with compassionate eyes she looked at him, and blessed him: 'Let you evolve as a great doctor, and continue to serve people'. She knew that she had to leave the phenomenal world. Yet, she could remain graceful, and her gentle words would charm everybody: 'Why should I worry? I am not alone. All of you are with me'.

What else does one need to live and die meaningfully?